DATA HANDBOOK
for
CLAY MATERIALS
and other
NON-METALLIC MINERALS

DATA HANDBOOK
for
CLAY MATERIALS
and other
NON-METALLIC MINERALS

Providing those involved in clay research and industrial
application with sets of authoritative data describing the
physical and chemical properties and mineralogical
composition of the available reference materials

Edited by

H. VAN OLPHEN

*Former Executive Secretary of the Numerical
Data Advisory Board of the National Academy of Sciences,
Washington DC, USA*

and

J. J. FRIPIAT

*Director of Research at the CNRS,
Professor at the University of Louvain, Past President of
the Association Internationale pour l'Etude des Argiles.
CRSOCI, Orleans, France*

Data prepared
under the auspices of the OECD
and the Clay Minerals Society

PERGAMON PRESS

OXFORD · NEW YORK · TORONTO · SYDNEY · PARIS · FRANKFURT

U.K.	Pergamon Press Ltd., Headington Hill Hall, Oxford OX3 0BW, England
U.S.A.	Pergamon Press Inc., Maxwell House, Fairview Park, Elmsford, New York 10523, U.S.A.
CANADA	Pergamon of Canada, Suite 104, 150 Consumers Road, Willowdale, Ontario M2J 1P9, Canada
AUSTRALIA	Pergamon Press (Aust.) Pty. Ltd., P.O. Box 544, Potts Point, N.S.W. 2011, Australia
FRANCE	Pergamon Press SARL, 24 rue des Ecoles, 75240 Paris, Cedex 05, France
FEDERAL REPUBLIC OF GERMANY	Pergamon Press GmbH, 6242 Kronberg-Taunus, Pferdstrasse 1, Federal Republic of Germany

First edition 1979

British Library Cataloguing in Publication Data

Data handbook for clay materials and other non-metallic minerals.
1. Clay minerals — Handbooks, manuals, etc.
2. Nonmetallic minerals — Handbooks, manuals, etc.
I. Van Olphen, H II. Fripiat, J III. Clay Minerals Society IV. Organisation for Economic Co-operation and Development
549'.67   QE389.625   78-41214
ISBN 0-08-022850-X

In order to make this volume available as economically and as rapidly as possible the authors' typescripts have been reproduced in their original forms. This method unfortunately has its typographical limitations but it is hoped that they in no way distract the reader.

42174

Printed and bound at William Clowes & Sons Limited Beccles and London

CONTENTS

ACKNOWLEDGEMENTS

CMS Project

   As agreed in organizing the project, each contribution is
identified with the participating author's name .

OECD Project

   Individual experimental results are not identified with the
participating laboratory (except by code number) or the authors in
accordance with the conditions accepted by the participants. A list
of the participating institutions is shown below, with the name of
the Director (D) and the participant-contact person (P). In most
laboratories several persons participated actively in the work, and
their combined anonymous efforts have been invaluable for the success
of the project.

   For each property studied, the results from the various laboratories
were collected and analyzed by "rapporteurs" whose names do
appear as the authors of the respective summaries.

Participating Institutes (°)
OECD Project

(D) Director   (P) Participant-Contact Person

Austria

E.Schroll (D)

Grundlageninstitut der Bundesversuchs-

   und Forschungsanstalt Arsenal,

   Vienna

(°) Names of Institutes and Directors and the locations are those applicable
   at the time the project was conducted.

Acknowledgements

Belgium

J.J.Fripiat (D)(P)
Laboratoire de Physico-Chimie
Minérale
  Heverlee-Louvain

J. de Cuyper (D)
Laboratoire de Traitement des
Minérais
  Heverlee-Louvain

Canada

H.M.Woodrooffe (D)
Mineral Processing Division
Dep. of Energy, Mines and Resources
Ottawa

France

M.Steinberg (P)
Laboratoire de Sédimentologie
Faculté des Sciences
Orsay

J.Gerard-Hirne (D)
J.Tuleff (P)
Institut de Céramique Française
Sèvres

V.Gabis (D,P)
Laboratoire de Géochimie
Faculté des Sciences
Orléans-la Source

M.Harispe (D)
J.Y.Jeanneau (P)
Société Française de Céramique
Paris

J.Wyart(D)
G.Sabatier (P)
Laboratoire de Minéralogie
  de la Faculté des Sciences
  de Paris,associé au C.N.R.S.

E.Plumat (D)
A.Jelli (P)
S.A.GLAVERBEL
Gilly

J.Wyart (D)
A.Oberlin
Laboratoire de Minéralogie-
  Cristallographie
Faculté des Sciences
Paris

J.Orcel (D)
S.Caillère
Laboratoire de Minéralogie du Muséum
  National d'Histoire Naturelle
Paris

S.Hénin (D)
J.Chaussidon (P)
Station Centrale d'Agrononomie
  C.N.R.A.
Versailles

G.Pedro
Laboratoire des Sols C.N.R.A.
Versailles

M.Roubault (D)
H.de la Roche (P)

Paris

G.Branche (D)
F.Chantret (P)
Service de Minéralogie
Direction des Productions
Chatillon-sous-Bagneux
R.Guennelon (D,P)
Station d'Agronomie
Domaine Sait-Paul
Montfavet

Germany BRD

K.Konopicky (D)
I.Patzak (P)
Forschungsinstitut der
  Feuerfest Industrie
Bonn

D.Schroeder (D)
Institut für Pflanzenernährung
  und Bodenkunde der Universität
Kiel

von Gaertner (D)
Eckhardt (P)
Harre (P)
Bundesanstalt für Bodenforschung
Hannover-Buchholz

U.Schwertmann (D,P)
Institut für Bodenkunde
  Technische Universität
Berlin

U.Hofmann (D,P)
Anorganisch-Chemisches Institut
  der Universität
Heidelberg

K.Jasmund (D,P)
Mineralogisch-Petrographisches
  Institut der Universität

J.Cases (P)
Centre de Recherches Pétrographiques
  et Géochimiques
Nancy

C.Guillemin (D)
Jacqueline Sarcia (P)
B.R.G.M.
Orléans-la Source

K.V.Gehlen (D)
H.Krumm (P)
Institut für Petrologie, Geochemie
  und Lagerstättenkunde der Universität
Frankfurt/Main

H.Kromer (D,P)
Abt. Sedimentpetrographie-Keramik
  Staatliches Forschungsinstitut
  für Angewandte Mineralogie
Regensburg

D.Huffmann (D,P)
Preussag AG
  Erdöl und Bohrverwaltung
  Geologisches Laboratorium
Berkhöpen/Peine

F.Karl (D)
P.Hörmann (P)
Mineralogisch-Petrographisches
  Institut der Universität
Kiel

H.Tributh (P)
Institut für Bodenkunde und
  Bodenerhaltung
Giessen

H.Harder (D)

Köln

D.Riedel (P)

Geologisches Institut der
   Ruhr Universität

Bochum

K.H.Papenfuiz (D,P)

Institut für Bodenkunde

Stuttgart

H.E.Schwiete (D)

W.Krönert(P)

Institut für Gesteinshüttenkunde
   der Rhein.-Westf. Technische
   Hochschule

Aachen

R.Hesse(P)

Institut für Geologie
   Technische Hochschule

München

W.Fleming (P)

A.Heydemann (P)

W.Smykatz-Kloss (P)

Sedimentpetrographisches Institut
   der Universität

Göttingen

P.Hahn-Weinheimer (D)

Forschungsstelle für Geochemie
   Institut für Mineralogie
   Technische Hochschule

München

P.Richter (D)

A.Peters (P)

Mineralogisches Institut der
   Universität

Würzburg

Greece

Th.Skoulikidis (D)

P.Korogiannakis (P)

Laboratoire de Chimie Physique et
   d'Electrochimie Appliquée
   Université Technique Nationale

Athens

N.Rakintzis (D)

Ch. Markopoulos (P)

"Democritos",Chemistry Division
   Greek Nuclear Resarch Centre

Athens

A.Kostikas (D,P)

"Democritos",Physics Division
   Greek Nuclear Research Centre

Athens

N.Yassoglou (D,P)

"Democritos", Soils and Plant
   Nutrition Laboratory
   Greek Nuclear Research Centre

Athens

Ireland

R.J.Nichol (D)

A.P.Carroll (P)

Inorganic Chemistry/Minerals
   Laboratory, Institute for
   Industrial Research and Standards

Dublin

Italy

P.Gallitelli (D,P)

Instituto di Mineralogia e
   Petrografia
   Università di Bologna

Bologna

Acknowledgements   xiii

Japan   Portugal

T.Ando (D)   M.Rocha (D)
Yasuyashi Watanabe (P)   A.Valeriana de Seabra (P)
Government Industrial Research   Laboratório Nacional de
    Institute of Osaka   Engenharia Civil
Osaka   Divisão de Química
                                              Lisbon

Sweden

U.Landergren (D)   P.G.Kihlstedt (D)
C.Helgesson (P)   E.Forssberg (P)
Materials Research Department   Mineral Processing Division
National Defense Research Inst.   Kungl.Tekniska Högskolan
Stockholm   Stockholm

Switzerland

R.Iberg (D)
Th.Mumenthaler
Laboratorium der Züricher
    Ziegeleien
Zürich

United Kingdom

A.Hodgson   N.F.Astbury (D)
Cape Asbestos Fibres,Ltd.   P.S.Keeling (P)
London   The British Ceramic Research
                                                  Association
K.C.Dunham , P.J.A.Bain   Stoke-on-Trent
Institute of Geological Sciences
London   V.J.Francis (D)

A.D. White   H.P.Rooksby (P)
J.H.Sharp   The General Electric Co.Ltd.
Dep. of Ceramics with Refractories   North Wembley,Middlesex
    Technology   D.Taylor (P)
University of Sheffield   Doulton Research Ltd.
Sheffield   Basil Green Laboratory
                                              Chertsey,Surrey
B.Lincoln (D)
L.J.Monkman (P)   D.Mitchell (D)
Turner and Newall Ltd.   A.Stott (P)
Asbestos Fibre Laboratory   Watts,Blake,Bearne and Co.Ltd.

Manchester

L.V.I.Berkin (D,P)
Pike Bros.,Fayle and Co.Ltd.
Wareham, Dorset

A.B.Stewart (D)
R.C.Mackenzie (P)
V.C.Farmer (P)
The Macaulay Institute for
   Soil Research
Craigiebuckler-Aberdeen

C.C.Hall (D)
A.S.Joy (P)
Warren Spring Laboratory
Stevenage, Herts.

M.G.Fleming (D)
A.P.Prosser(P)
M.P.Jones (P)
Imperial College of Science
   and Technology, Dept. of
   Mining and Mineral Technology
   Royal School of Mines
London

Newton Abbot, Devon

N.O.Clark (D ,P)
Central Research Laboratories
English China Clays Ltd.
St.Austell, Cornwall

W.L.German (D.P)
North Staffordshire College
 of Technology
Stoke-on-Trent

F.S.Spring (D)
B.S.Neumann (P)
Laporte Industries Ltd.
   O. and P. Division
Redhill,Surrey

INTRODUCTION

Background

Engineers, geologists, mineralogists and others involved in clay
research have always felt the need for well characterized samples
of typical clays. The general use of such "reference clays" in
research enables a ready comparison of published results, and in
the course of time a wealth of qualitative and quantitative data
will thus be added to the information which was originally collected
for the characterization of the samples. The reference materials can
also be used for comparing and evaluating test methods and laboratory
performance.

The first project intended to meet this need was initiated in the
late forties under the auspices of the American Petroleum Institute
and Columbia University. Batches of typical clays, called "Reference
Clay Minerals" were collected and homogenized.Property data on these
materials were determined in a number of industrial and university
laboratories on a voluntary basis. The results were published in a
series of eight "American Petroleum Institute - Research Project 49
Reports" by Columbia University, New York. The samples were made
available through Ward's? The Reference Clay Minerals were widely used
in research. When supplies of the original batches were exhausted,
some new batches were procured from the same general locations, but
these were no longer identical with those on which the data had been
obtained. Also, some of the original data became obsolete with the

----

Ward's Natural Science Establishment, Inc., Rochester,N.Y, and
    Monterey,Cal.,USA

1

introduction of better or more standardized experimental methods.

In 1967 a new effort was initiated under the auspices of the OECD
as part of a project called "Research cooperation on non-metallic
minerals", in which laboratories in European member countries and Japan
participated (1). A few years later another analogous project was
launched in the U.S.A. under the auspices of the Clay Minerals Society,
called the "Source Clays Program"(2). In this project, both U.S.
and European laboratories participated. A feature of the U.S.
project is the large size of the batches - about 1000 lbs each -
which allows the distribution of samples of one pound each, whereas
the samples made available in the OECD project is in the gram range
out of batches of 20-40 pounds. Instructions for ordering the samples
are given below.

Since the OECD discontinued its materials program including the
project on clays and other non-metallic minerals at the time when
only a summary of the results had been prepared (1), complete results
from the individual laboratories have not been published before. When
the Clay Minerals Society considered alternatives for the publication
of the data collected on its suite of samples, it was decided to
combine the data on the CMS and OECD projects in the present data
handbook.

S c o p e

The samples were carefully removed at the sites, pulverized and
homogenized. Although they consist primarily of a single clay
mineral, they contain various mineral impurities commonly associated
with the clay minerals. Therefore,the samples should be considered
typical <u>clay materials</u> rather than pure clay minerals. However, the
materials will be convenient starting materials for the preparation
of pure clay minerals. As long as the purification procedures are
adequately described, the purified minerals may then in turn be considered reference clay minerals, for which data may be compared
in the literature.

The data presented in this handbook apply generally to the
materials as distributed, or, in some cases to samples which were
subjected to some described pretreatment. The data are, therefore,
not exactly characteristic for the clay mineral itself, but close
to it as long as interference from impurities is small. <u>The user</u>
<u>of the data should give careful attention to amount and kind of</u>

<u>impurities as reported when making judgements regarding the applic-</u>
<u>ability of the data for the characterization of the clay mineral</u>
<u>which is the main constituent.</u> In some cases the amount of impur-
ities is substantial!

There are some differences in emphasis in the two projects. For
example, for the CMS suite of samples, the location is described
in greater detail, primarily for the geologist user of the data.
The OECD project which was conducted on a larger scale aimed at
a statistical evaluation of the results, therefore, property data
on each sample were solicited from a large number of laboratories.

S e l e c t i o n   o f   s a m p l e s

The sample collections were composed to contain representatives
of the main classes of clay minerals, i.e. smectites, kaolinites,
illites and attapulgite (palygorskite), depending on the avail-
ability of typical materials in sufficiently large quantities.
Synthetic materials are included. In addition, the CMS project
makes available a number of "special clays" which were carefully
collected in relatively small quantities and not processed in any
way. No data are presented on this group of materials in this book.
The OECD suite of samples comprises a number of non-clay materials
which constitute a regular part of the collection of non-metallic
minerals. Data for these materials are included in this handbook.

The Standard Reference Materials collection of the National
Bureau of Standards in the U.S.A. contains two clays for which
a certified chemical analysis is provided. These are useful for
checking methods of chemical analysis.

Names and code numbers of the samples in these collections
are given in the Tables below, together with instructions for
ordering samples.

R e s u l t s

The experimental work on these projects has been carried out
exclusively on a voluntary basis. Consequently, it often has been
impossible to realize measurement programs in a way to satisfy
rigorous statistical requirements.Hence, the evaluation of random
errors to assess the imprecision of the results was often less
than ideal, and judgements on systematic errors, and hence the
overall accuracy,were made arbitrarily, if made at all. Based
on the experience gained, a second phase of improved efforts

was projected for the OECD samples, but unfortunately this second
round could not be implemented.

The general experience with projects of this kind is that
data obtained by different laboratories on the same sample
show a considerable spread, even in spite of many precautions
taken to obtain comparable results. The results of the projects
presented in this handbook are no exception. This demonstrates
once more the need for careful attention to experimental detail,
for the standardization of experimental procedures, and the need
for certified reference materials with which individual procedures
may be evaluated. Also, in many cases, experimental details were
not reported in a way to allow the expert reader a proper
evaluation of the reported data. This experience demonstrates
the need for basic rules for reporting experimental data in
the literature. Recommendations for authors and editors are
formulated for a growing number of disciplines under the auspices
of CODATA, the Committee on Data for Science and Technology of
the International Council of Scientific Unions, and several
Unions of ICSU and other international scientific societies(3).

In spite of such imperfections of the projects, many very
excellent data have been collected which will be helpful for
those engaged in research and development efforts involving these
materials. In making the materials available, it is anticipated
that the users will publish many additional data in the open
literature. Authors are urged to describe any pretreatment procedures properly since some of the material properties are
highly sensitive to variations in sample preparation.

The entire project has been rather educational for the participants. In order to share the gained experience, one of the
participants will discuss the various methodological aspects
for each of the properties studied, outlining the merits and
limitations of the applied methods and the validity of the conclusions drawn. For a full description of measurement methods,
however, the reader is referred to the literature.

Organization of the handbook

In Part I the data are summarized and presented by sample.
In this part the sample location is described and a brief
mineralogical characterization, derived from the results of
all applied methods, is given to indicate the kind and estimated

amounts of impurities present in the samples.

In Part II the data are presented in detail, arranged by the property measured. Although the applied methods are not described as in a laboratory manual, references are given and recommendations are made regarding preferred methods, sample treatment, procedure, instrument calibration, and reporting of results, whenever desirable.

R e f e r e n c e s

(1) Fripiat,J.J., Minéraux non-métalliques, Final Summary, OECD, 16 November, 1970. Limited Distribution.

(2) Moll,W.F.Jr., Johns,W.D., and van Olphen,H., The Source Clays Program. Proc.International Clay Conference 1975. (S.W.Bailey, Editor) Allied Publishing Ltd.,Wilmette, Illinois, 1976

(3) Guide for the presentation in the primary literature of numerical data derived from experiments. CODATA Bulletin No. 9, December 1973 6 pp.

(Also distributed by Unesco in several translations) Contains references to a series of specialized guides for various disciplines and sub-disciplines.

Data Handbook
CLAY MINERALS SOCIETY
"SOURCE CLAYS"

code	name	supplier
KGa-1	Kaolinite, Georgia, well crystallized	Georgia Kaolin Company
KGa-2	Kaolinite, Georgia, poorly crystallized	Georgia Kaolin Company
SWy-1	Montmorillonite, Wyoming	Baroid Division, NL Industries
STx-1	Montmorillonite, Texas	Southern Clay Products Company
SAz-1	Montmorillonite, Arizona (Cheto)	Filtrol Corporation
SHCa-1	Hectorite, California	Baroid Division, NL Industries
Syn-1	Synthetic mica-montmorillonite Barasym SSM-100 $^{(R)}$	Baroid Division, NL Industries
PFl-1	Attapulgite, Florida	Engelhard Minerals and Chemicals Co.

1000 lb batches were homogenized by Baroid Division, NL Industries

"SPECIAL CLAYS"

SWa-1   Ferruginous Smectite, from Washington State
      collected by J.A.Kittrick, Washington State University, Pullman

SCu-1   Otay Montmorillonite, from California
      collected by J.L.Post, California State University, Sacramento

RAr-1   Rectorite, a mixed layer mineral, from Arkansas
      collected by C.Stone, Arkansas Geological Commission

CAr-1   Cookeite, a chlorite, from Arkansas
      Collected by C.Stone, Arkansas Geological Commission

.....   Nontronite, from Washington State

The samples have been listed in "Reference Samples for the Earth Sciences" by F.J.Flanagan, Geochimica et Cosmochimica Acta, 38,1731,1974

ORDERS

Orders should be directed to the curator of the sample collection: Prof.William D.Johns, Department of Geology, University of Missouri, Columbia, Mo 65201, U.S.A.

The handling costs for each source clay sample is presently $10.00 per 500 g, post paid by surface mail. Air postage is additional. Checks or money orders should accompany orders and should be made out to: "U. of Mo. Curators-Source Clays", and should include estimated air mail postage, if air mail service is desired.

The unit sample size for the special clays is variable and samples should be ordered separately, indicating desired amounts.

OECD BANK SAMPLES

No.	Name Origin	Supplier
01	Montmorillonite, Camp Berteau,Morocco	S.Caillère,Laboratoire de Minéralogie du Muséum National d'Histoire Naturelle,Paris,France
02	Laponite, synthetic hectorite	F.S.Spring,Laporte Industries,Ltd. Redhill,Surrey, U.K.
03	Kaolinite (China Clay),St.Austell,U.K.	N.O.Clark,English China Clays,Ltd. St.Austell, Cornwall, U.K.
04	Attapulgite, Attapulgis, Florida, U.S.	K.C.Dunham, Institute of Geological Sciences, London, U.K.
05	Illite, Le Puy en Velay, France	V.Gabis, Laboratoire de Géochimie, Faculté des Sciences, Orléans-la Source
06	Chrysotile (Asbestos,Cassiar)	H.M.Woodrooffe, Mineral Processing Division, Dep. of Energy, Mines and Resources, Ottawa, Canada
07	Crocidolite, Koegas Mine, Province du Cap, South Africa	A.Hodgson, Cape Asbestos Fibres, London, U.K.
08	Talc ,Valle del Chisone, Piemont, Italy	P.Gallitelli, Institute of Mineralogy, University of Bologna, Italy
10	Gibbsite, synthetic	H.E.Schwiete, Institut für Gesteinshüttenkunde der Rhein.-Westf. Techn. Hochschule,Aachen,BRD
11	Magnesite , Austria	E.Schroll, Grundlageninstitut der Bundesversuchs-und Forschungsanstalt Arsenal, Vienna, Austria
12	Calcite , Ireland	R.J.Nichol, Inorganic Chemistry Institute for Industrial Research, Dublin, Ireland
13	Gypsum, Carrière de Vaujours, France	Cl.Guillemin, B.R.G.M., Orléans-la Source

20-40 kg batches were homogenized by Warren Spring Laboratory,
    Stevenage, Herts., U.K.

Samples may be requested by writing to:

Mlle. S.Caillère, Laboratoire de Minéralogie du Muséum National
    d'Histoire Naturelle, 61, rue de Buffon, Paris 5e, France

STANDARD REFERENCE MATERIALS

Issued by the National Bureau of Standards

This Catalog lists and describes the Standard Reference Materials (SRM's), Research Materials (RM's), and General Materials (GM's) currently distributed by the National Bureau of Standards, as well as many of the materials currently in preparation. SRM's are used to calibrate measurement systems and to provide a central basis for uniformity and accuracy of measurement. The unit and quantity, the type, and the certified characterization are listed for each SRM, as well as directions for ordering. The RM's are not certified, but are issued to meet the needs of scientists engaged in materials research. RM's are issued with a "Report of Investigation," the sole authority of which is the author of the report. The GM's are standardized by some agency other than NBS. NBS acts only as a distribution point and does not participate in the standardization of these materials. Announcements of new and renewal SRM's, RM's, and GM's are made in the semi-annual supplements to this Catalog, SRM Price List, and in scientific and trade journals.

Key words: Analysis; characterization; composition; properties; Standard Reference Materials; Research Materials; General Materials.

General Information

All of the Standard Reference Materials (SRM's), Research Materials (RM's), and General Materials (GM's) listed in this Catalog bear distinguishing names and numbers by which they are permanently identified. Each SRM, RM, or GM bearing a given designation is of identical characterization with every other sample bearing the same designation, within the limits required by the use for which it is intended; or if necessary, it is given a serial number and an individual calibration.

The first SRM's issued by the Bureau were a group of ores, irons, and steels certified for chemical composition, and by custom they came to be called "standard samples." At present, nearly 900 SRM's are available, covering a wide range of chemical and physical properties, and the designation, Standard Reference Material, is more appropriate. As the number of SRM's has increased, so has the variety, with such new groups being established as: clinical laboratory standards, nuclear materials, glass viscosity standards, rubber and rubber compounding materials, color standards, and coating thickness standards. These groups are listed under the headings: Standards of Certified Chemical Composition, Standards of Certified Physical Properties, Engineering Type Standards, Research Materials, or General Materials. The groups of materials under these general headings are listed in the Table of Contents. An alphabetical index provides the location of a particular material, or group of similar materials. A numerical index provides the date of the current Certificate issued with these materials.

The detailed listing of materials indicates the nominal certification for which the material is issued, but the Certificate must be consulted for the actual certification A number of SRM's are issued for which it is not feasible to supply numerical values, or for which such certification would not be useful. These SRM's provide assurance of identity among all samples with the same designation, and permit standardization of test procedures and referral of physical or chemical data on unknown materials to a common basis.

Preparation and Availability of Standard Reference Materials

New SRM's are prepared each year and are announced through periodic supplements to this Catalog as well as directly, to prospective users.

The preparation of "renewal" SRM's is intended to be completed by the time the existing supply of each kind of material is exhausted, but this is not always possible. The renewal will not usually be identical to its predecessor, but will be quite similar especially with regard to the characteristics certified, and generally the renewal can be used in place of its predecessor. As an example, when the first 0.1 percent carbon Bessemer steel was prepared in 1909, it was called Standard Sample No. 8. During the following years, a number of renewal batches, 8a, 8b, etc., were prepared; SRM 8j is now available and represents the 10th renewal batch of 0.1 percent carbon Bessemer steel. While each of these batches differ somewhat in detailed analysis from one batch to another, all retain the relatively high level of phosphorus, sulfur, and nitrogen, and low alloy metal content characteristic of this type of material. It is not possible to supply preceding numbers of a renewal series when the stock is exhausted. If little demand exists or an alternate source of supply has become available for a material, production may be discontinued permanently or until sufficient justification is obtained to warrant renewal.

Supplements to the Catalog are issued periodically to keep it current. These supplements provide a complete list of the available SRM's and their prices and provide descriptions of SRM's issued since the latest Catalog was printed.

Ordering

Orders should be addressed to the Office of Standard Reference Materials, Room B311, Chemistry Building, National Bureau of Standards, Washington, D.C. 20234. Telephone (301) 921-2045. Orders should give the amount (number of units), catalog number and name of the standard requested. For example: 1 each, No. 11h, Basic-Open-Hearth Steel, 0.2 percent C. These materials are distributed only in the units listed.

Acceptance of an order does not imply acceptance of any provision set forth in the order contrary to the policy, practice, or regulations of the National Bureau of Standards or the U.S. Government.

Orders received for "out-of-stock" materials are cancelled if only out-of-stock items are ordered. On other orders, shipment is made of available materials and out-of-stock items are cancelled. Back-orders are not accepted for out-of-stock materials; if a renewal lot of material is available, it will be furnished automatically.

Terms

Prices are given in the SRM Price List. These prices are subject to revision and orders will be billed for prices in effect at the time of shipment. New SRM Price Lists, when issued, are sent to users who have made purchases during the preceding twelve months, and to persons or organizations who request them. No discounts are given on purchases of Standard Reference Materials.

Remittances of the purchase price need not accompany purchase orders. Payment of invoices is expected within 30 days of receipt of an invoice. Payment on foreign orders may be made by any of the following:

(a) UNESCO coupons,
(b) banker's draft against U.S.A. bank,
(c) bank to bank transfer to a U.S.A. bank,
(d) letter of credit* on a U.S.A. bank, or
(e) by International Money Order.

Pro-forma invoice service will frequently require 6 to 8 weeks to process, and will be furnished only to those requiring such service.

*Letters of Credit may be used as advance payment for SRM's. Letters of credit will be accepted from banks

in the United States only. Listed below are the only documents that we will furnish:

(1) Six Commercial Invoices
(2) Packing List
(3) Certificate of Origin
(4) Airway Bill (only if material is shipped Collect).

If we ship material (*Prepaid*) International Air Parcel Post, we cannot furnish receipt.

Domestic Shipments

Shipments of material (except for certain restricted categories, e.g., hydrocarbons, special nuclear materials, compressed gases, rubber, rubber compounding materials, and radioactive standards) intended for the United States, Mexico, and Canada are normally shipped prepaid (providing that the parcel does not exceed the weight limitations as prescribed by Postal Laws and Regulations) unless the purchaser requests a different mode of shipment, in which case the shipment will be sent collect. The Bureau does not prepay such shipping charges. Hydrocarbons, organic sulfur compounds, compressed gases, rubber, rubber compounding materials, radioactive standards, and similar materials are shipped express collect.

Foreign Shipments

Orders for small weight shipments will be shipped by prepaid International Air Parcel Post. Other shipments will be shipped prepaid International Parcel Post, except those shipments exceeding the parcel post weight limitations, which must be handled through an agent (shipping or brokerage firm) located in the U.S.A. as designated by the purchaser. Shipments handled through an agent will be packed for overseas shipment and forwarded via express collect to the U.S.A. firm designated as agent.

NOTE: Orders and inquiries submitted in English will be processed more rapidly than those requiring translations.

NATIONAL BUREAU OF STANDARDS , USA
STANDARD REFERENCE MATERIALS (SRM)

Minerals

Chemical Composition
(Nominal Weight Percent as the Oxide)

SRM	Type	Wt/Unit (grams)	SiO$_2$	Fe$_2$O$_3$	Al$_2$O$_3$	TiO$_2$	MnO	CaO
1b	Limestone, argillaceous	50	4.92	0.75	1.12	0.046	0.20	50.9
88a	Limestone, dolomitic	50	1.20	.28	0.19	.02	.03	30.1
70a	Feldspar, potash	40	67.1	.075	17.9	.01	----	0.11
99a	Feldspar, soda	40	65.2	.065	20.5	.007	----	2.14
97a	Clay, flint	60	43.7	.45	38.8	1.90	----	0.11
98a	Clay, plastic	60	48.9	1.34	33.2	1.61	----	.31
81a	Glass sand	IN PREP						
165a	Glass sand (low iron)	IN PREP						
154b	Titanium dioxide	90	----	----	----	99.74	----	----

SRM	Na$_2$O	K$_2$O	Li$_2$O	ZrO$_2$	BaO	Rb$_2$O	P$_2$O$_5$	CO$_2$	Loss on Ignition	Cr$_2$O$_3$	MgO	SrO
1b	0.04	0.25	----	----	----	----	0.08	40.4	41.1	----	0.36	0.14
88a	.01	.12	----	----	----	----	.01	46.6	46.7	----	21.3	.01
70a	2.55	11.8	----	----	.02	.06	----	----	0.40	----	0.02	----
99a	6.2	5.2	----	----	.26	----	.02	----	0.26	----	.15	.18
97a	.037	.50	.11	.063	.078	----	.36	----	13.32	.03	----	----
98a	.082	1.04	.070	.042	.03	----	.11	----	12.44	.03	.42	.039
81a												
165a												
154b												

PART 1

DATA SUMMARIES
CLAY MINERALS SOCIETY

KGa-1 KAOLINITE
well crystallized

ORIGIN Tuscaloosa formation? (Cretaceous?) (stratigraphy uncertain)
County of Washington, State of Georgia, USA.

Location: 32°58' N - 82°53' W approximately, topographic map
  Tabernacle, Georgia N 3252.5 - W 8252.5/7.5
  Collected from face of Coss-Hodges pit, 3 October, 1972

CHEMICAL COMPOSITION (%)

SiO_2: 44.2 Al_2O_3: 39.7 TiO_2: 1.39 Fe_2O_3: 0.13 FeO: 0.08
MnO: 0.002 MgO: 0.03 CaO: n.d. Na_2O: 0.013 K_2O: 0.050 F: 0.013
P_2O_5: 0.034 Loss on heating: -550°C: 12.6; 550-1000°C: 1.18

CATION EXCHANGE CAPACITY
2.0 meq/100g

SURFACE AREA
N_2 area: 10.05 ± 0.02 m^2/g

THERMAL ANALYSIS
DTA: endotherm at 630°C, exotherm at 1015°C
TG: dehydroxylation weight loss 13.11% (theory 14%)
    indicating less than 7% impurities.

INFRARED SPECTROSCOPY
Typical spectrum for well crystallized kaolinite, however not
as well crystallized as a typical China Clay from Cornwall,
as judged from the intensity of the 3669 cm^{-1} band. Splitting
of the 1100 cm^{-1} band is due to the presence of coarse crystals.

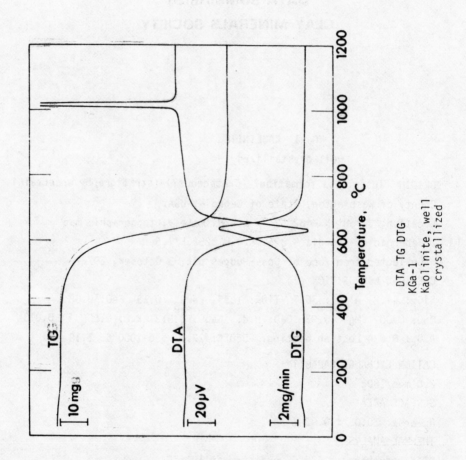

DTA TG DTG
KGa-1
Kaolinite, well
crystallized

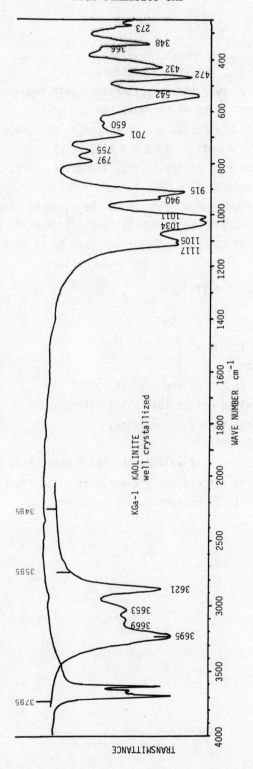

KGa-1 KAOLINITE
well crystallized

KGa-2   KAOLINITE
poorly crystallized

ORIGIN Probably lower tertiary (stratigraphic sequence uncertain)

County of Warren, State of Georgia, USA

Location: 33°19' N - 82°28' W approximately, topographic map

  Bowdens Pond, Georgia N 3315 - W 8222.5/7.5

  Collected from face of Purvis pit, 4 October, 1972

CHEMICAL COMPOSITION (%)

SiO_2: 43.9 Al_2O_3: 38.5 TiO_2: 2.08 Fe_2O_3: 0.98 FeO: 0.15

MnO: n.d. MgO: 0.03 CaO: n.d. Na_2O:<0.005 K_2O: 0.065

P_2O_5: 0.045 S: 0.02 Loss on heating: -550°C: 12.6; 550-1000°C:

1.17 F: 0.020

CATION EXCHANGE CAPACITY

3.3 meq/100g

SURFACE AREA

N_2 area: 23.50 ± 0.06 m^2/g

THERMAL ANALYSIS

DTA: endotherm at 625°C, exotherm at 1005°C

TG: dehydroxylation weight loss 13.14% (theory 14%)

  indicating less than 7% impurities.

INFRARED SPECTROSCOPY

Typical spectrum of less well crystallized kaolinite, however

  the mineral is not extremely disordered since the band at

  3669 cm^{-1} is still present in the spectrum.

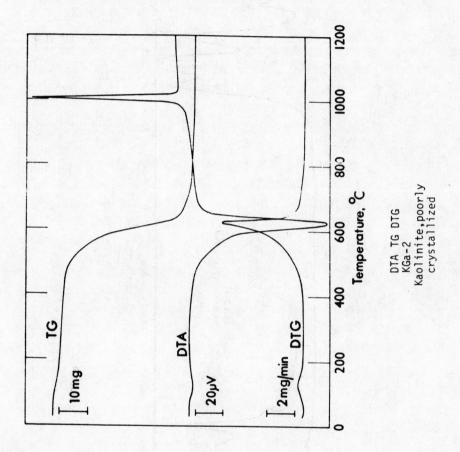

DTA TG DTG
KGa-2
Kaolinite,poorly
crystallized

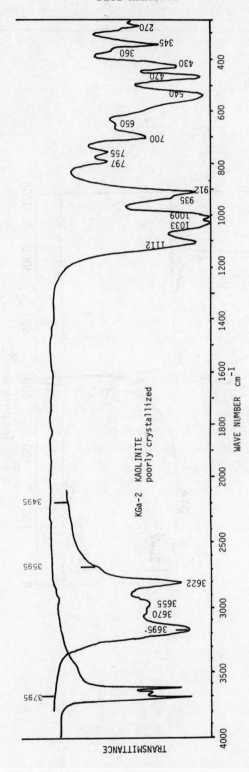

KAOLINITE
KGa-2 poorly crystallized

SWy-1 MONTMORILLONITE WYOMING

ORIGIN Newcastle formation (cretaceous)
Crook County, State of Wyoming,
Location: NE 1/4 SE 1/4 Sec.18, T 57 N, R 65 W; topographic
  map: Seeley (15'); The upper 63 of recently stripped area
  was removed to expose clean, green upper Newcastle, from
  which sample was taken, 3 October, 1972
CHEMICAL COMPOSITION (%)
SiO_2: 62.9 Al_2O_3: 19.6 TiO_2: 0.090 Fe_2O_3: 3.35 FeO: 0.32
MnO: 0.006 MgO: 3.05 CaO: 1.68 Na_2O: 1.53 K_2O: 0.53
P_2O_5: 0.049 S: 0.05 F: 0.111 Loss on heating: -550°C: 1.59;
550-1000°C: 4.47; CO_2: 1.33

CATION EXCHANGE CAPACITY
76.4 meq/100g, principal exchange cations Na and Ca
SURFACE AREA
N_2 area: 31.82 ± 0.22 m^2/g
THERMAL ANALYSIS
DTA: endotherms at 185°C (shoulder at 235°C), desorption of
water; 755°C, dehydroxylation; shoulder at 810°C; exotherm
at 980°C.
TG: Loss in dehydroxylation range : 5.53% (theory 5.0%)
INFRARED SPECTROSCOPY
Typical spectrum for Wyoming bentonite with a moderate Fe^{3+}
content (band at 885 cm^{-1}). Quartz is detectable (bands
at 780, 800, 698, 400, and 373 cm^{-1}), and a trace of carbonate
(band at 1425 cm^{-1}).

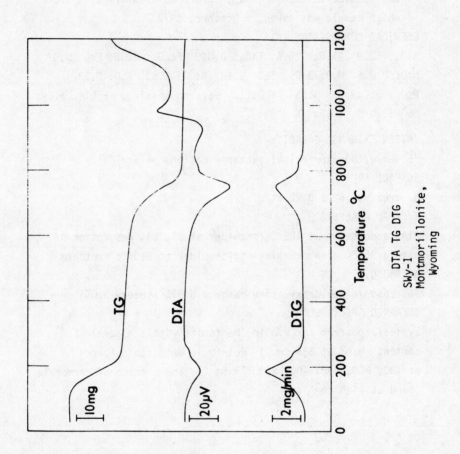

DTA TG DTG
SWy-1
Montmorillonite,
Wyoming

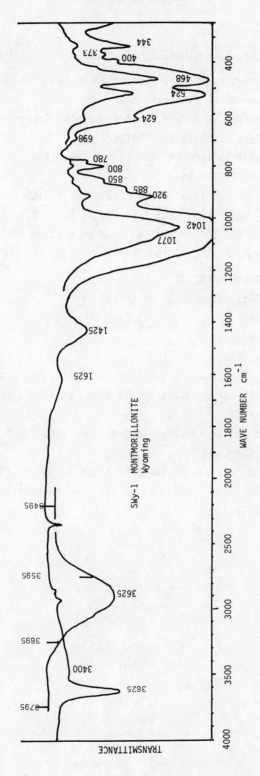

SWy-1 MONTMORILLONITE
Wyoming

STx-1 MONTMORILLONITE, TEXAS

ORIGIN Manning formation, Jackson group (eocene)
County of Gonzales, State of Texas
Location: 29°30' N, 97°22' W approximately. Topographic map
    Hamon, Texas, N 2922.5 - W 9715/7.5
    Collected from face of pit, 17 October, 1972

CHEMICAL COMPOSITION (%)

SiO_2: 70.1 Al_2O_3: 16.0 TiO_2: 0.22 Fe_2O_3: 0.65 FeO: 0.15
MnO: 0.009 MgO: 3.69 CaO: 1.59 Na_2O: 0.27 K_2O: 0.078
P_2O_5: 0.026 S: 0.04 F: 0.084 Loss on heating -550°C: 3.32;
550-1000°C: 3.22 CO_2: 0.16

CATION EXCHANGE CAPACITY
84.4 meq/100g, major exchange cation Ca

SURFACE AREA
83.79 $\pm$ 0.22 m^2/g

THERMAL ANALYSIS
DTA: endotherms at 185°C (shoulder at 235°C), desorption of
water; 720°C, dehydroxylation; shoulder at 920°C; exotherms
at 1055°C, 1065°C, 1135°C.

TG: Loss in dehydroxylation range: 3.88% (theory: 5.0%)

INFRARED SPECTROSCOPY
The spectrum indicates a low iron content. Quartz (697 cm^{-1}),
a silica phase (797 cm^{-1}), and a trace of carbonate (1400 cm^{-1})
are detectable.

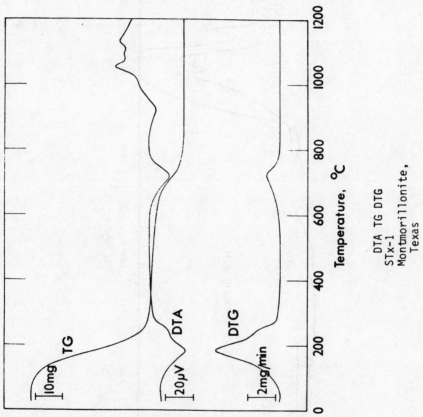

DTA TG DTG
STx-1
Montmorillonite,
Texas

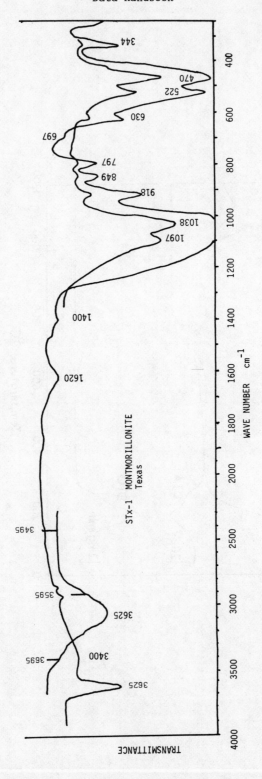

STx-1 MONTMORILLONITE
Texas

SAz-1 MONTMORILLONITE ARIZONA (CHETO)

ORIGIN Bidahochi formation (pliocene)

Apache County, State of Arizona.

Location: SE 1/4 NE 1/4 Sec. 26 T 21 N? R 29 E; topographic
   map: Gallup (1 : 250,000). Collected from pit after overburden was
   stripped, 8 May, 1973

CHEMICAL COMPOSITION (%)

SiO_2: 60.4 Al_2O_3: 17.6 TiO_2: 0.24 Fe_2O_3: 1.42 FeO: 0.08
MnO: 0.099 MgO: 6.46 CaO: 2.82 Na_2O: 0.063 K_2O: 0.19
P_2O_5: 0.020 F: 0.287 Loss on heating -550°C: 7.54; 550-
1000°C: 2.37

CATION EXCHANGE CAPACITY

120 meq/100g, major exchange cation Ca.

SURFACE AREA

N_2 area: 97.42 $\pm$ 0.58 m^2/g

THERMAL ANALYSIS

DTA: endotherms at 200°C, shoulder at 240°C, desorption of water;
685°C, dehydroxylation; shoulder at 895°C; exotherms at 1020°C,
1065°C, 1160°C

TG: Loss in dehydroxylation range: 4.69% (theory: 5.0%)

INFRARED SPECTROSCOPY

The spectrum indicates a low octahedral iron content. A silica
phase (band at 790 cm^{-1}) is detectable.

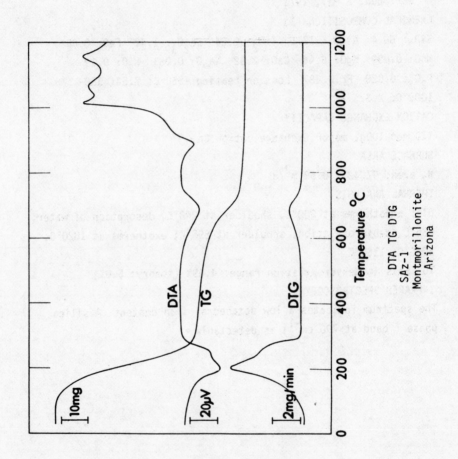

DTA TG DTG
SAz-1
Montmorillonite
Arizona

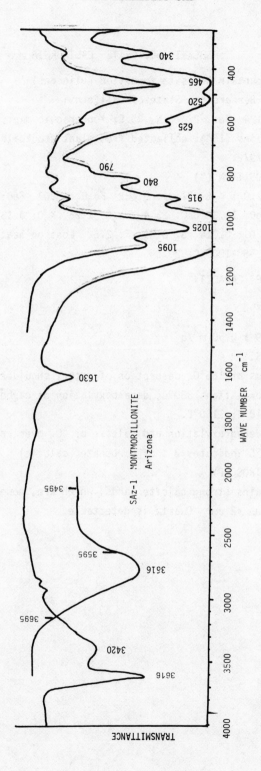

SAz-1 MONTMORILLONITE
Arizona

WAVE NUMBER cm⁻¹

SHCa-1 HECTORITE CALIFORNIA

ORIGIN Red Mountain Andesite formation (pliocene)
County of San Bernardino, State of California
Location: NE 1/4 Sec. 27 T8 N, R5 E; topographic map:
  Cady Mountains (15'); collected from plant stockpile,
  November, 1972.

CHEMICAL COMPOSITION (%)

SiO_2: 34.7 Al_2O_3: 0.69 TiO_2: 0.038 Fe_2O_3: 0.02 FeO: 0.25
MnO: 0.008 MgO: 15.3 CaO: 23.4 Na_2O: 1.26 K_2O: 0.13
Li_2O: 2.18 P_2O_5: 0.014 S: 0.01 F: 2.60 Loss on heating
-550°C: 1.20; 550-1000°C: 20.6

CATION EXCHANGE CAPACITY
43.9 meq/100g

SURFACE AREA
N_2 area: 63.19 ± 0.50 m^2/g

THERMAL ANALYSIS
DTA: endotherms at 165°C, desorption of water; shoulder at 725°C;
795°C, dehydroxylation; 880°C, decarboxylation of carbonate;
shoulder at 910°C; 1130°C.

TG:Ranges of dehydroxylation and release of CO_2 overlap; loss of
CO_2 above 810°C indicates 27% of carbonate (calcite)

INFRARED SPECTROSCOPY
Spectrum contains strong calcite bands, which are, however, absent
in the fraction <2 μm. Quartz is detectable.

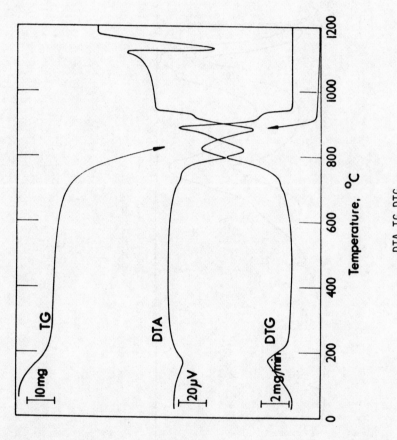

DTA TG DTG
SHCa-1
Hectorite
California

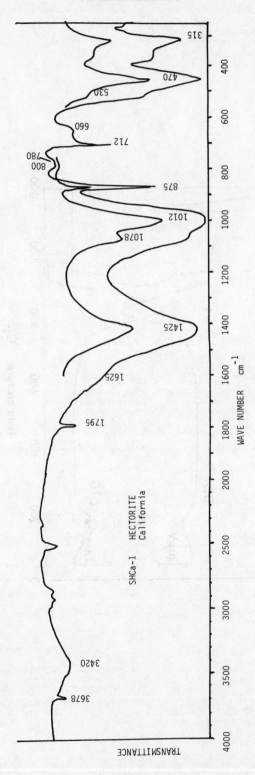

SHCa-1   HECTORITE
          California

315, 470, 530, 660, 712, 780, 800, 875, 1012, 1078, 1425, 1625, 1795, 3420, 3678

WAVE NUMBER   cm^{-1}

TRANSMITTANCE

Syn-1   SYNTHETIC MICA-MONTMORILLONITE

ORIGIN Synthetic, trade name Barasym SSM-100

Baroid Division, NL Industries, date of manufacture: 1972

CHEMICAL COMPOSITION (%)

SiO_2: 49.7 Al_2O_3: 38.2 TiO_2: 0.023 Fe_2O_3: 0.02 MgO: 0.014
Na_2O: 0.26 K_2O:<0.01 Li_2O: 0.25 P_2O_5: 0.001 S: 0.10
F: 0.76 Loss on heating -550°C: 8.75; 550-1000°C: 2.40.

CATION EXCHANGE CAPACITY

Barium method ca. 70 meq/100g; ammonium method ca. 140 meq/100g

SURFACE AREA

N_2 area: 133.66 ± 0.72 m^2/g

THERMAL ANALYSIS

DTA: endotherms at 140°C, desorption of water; 575°C, dehydroxylation;
exotherm at 1030°C

TG: The weight loss in the dehydroxylation range is 10.35% due to
simultaneous loss of ammonium which is the major exchange cation.

INFRARED SPECTROSCOPY

The spectrum is broadly similar to that of muscovite and contains
bands due to NH_4 (1432 and 1404 cm^{-1}) and to NH_4Br from the reaction
in the KBr disk.

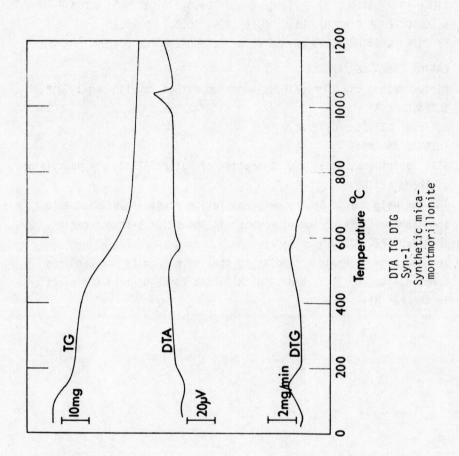

DTA TG DTG
Syn-1
Synthetic mica-
montmorillonite

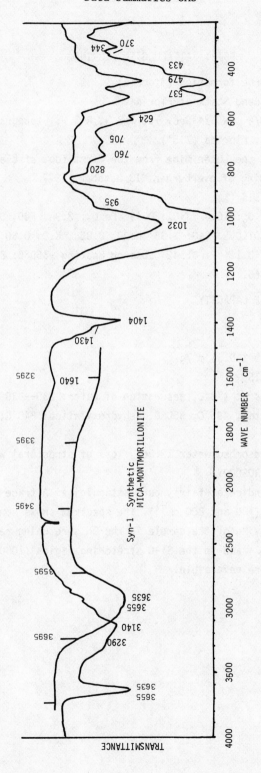

Syn-1   Synthetic
MICA-MONTMORILLONITE

PF1-1 ATTAPULGITE FLORIDA

ORIGIN Hawthorne formation (miocene)

County of Gadsden, State of Florida

Location: SE 1/4 NW 1/4 Sec. 10, T 3 N, R 3 W., topographic
  map: Dogtown, Florida (7.5')

  Collected at the Luten mine from the first foot of clay bed
  after stripping of overburden, 13 October, 1972

CHEMICAL ANALYSIS (%)

SiO_2: 60.9 Al_2O_3: 10.4 TiO_2: 0.49 Fe_2O_3: 2.98 FeO; 0.40
MnO: 0.058 MgO:10.2 CaO: 1.98 Na_2O: 0.058 K_2O: 0.80
P_2O_5: 0.80 S: 0.11 F: 0.542 Loss on heating -550°C: 8.66;
550-1000°C: 1.65.

CATION EXCHANGE CAPACITY

19.5 meq/100g

SURFACE AREA

N_2 area: 136.35 ± 0.31 m^2/g

THERMAL ANALYSIS

DTA: endotherms at 170°C, desorption of water; 230---300, desorption
of adsorbed water; 495°C; 550°C, dehydroxylation; 840°C; exotherm at
905°C

TG: Loss of adsorbed water 12.96%, loss of structural water 5.52%

INFRARED SPECTROSCOPY

The spectrum indicates fairly pure attapulgite. A trace of quartz
is detectable (780 and 800 cm^{-1}). The spectrum shows considerable
shifts upon drying of the sample in the OH stretching region
(3000-3700 cm^{-1}) and in the Si-O stretching region (1000-1200 cm^{-1})
These shifts are reversible.

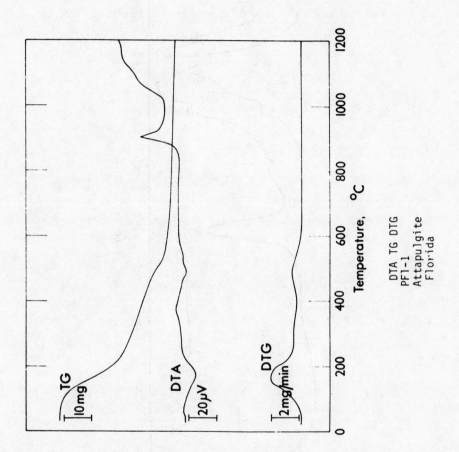

DTA TG DTG
PF1-1
Attapulgite
Florida

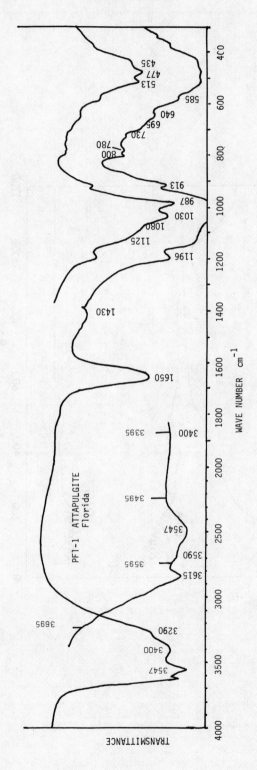

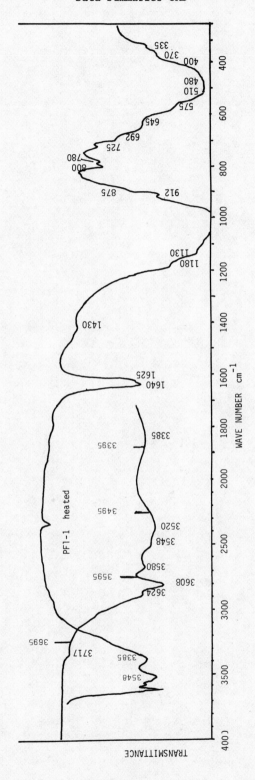

DATA SUMMARIES

ORGANIZATION FOR EUROPEAN COOPERATION AND DEVELOPMENT

01   MONTMORILLONITE

ORIGIN Camp Berteau, Morocco

CHEMICAL COMPOSITION (%)

SiO_2: 63.84 Al_2O_3: 22.24 TiO_2: 0.50 Fe_2O_3: 3.45 MnO: 0.01

MgO: 4.87 CaO: 2.88 Na_2O: 1.73 K_2O: 0.40 P_2O_5: 0.08 SO_3: 0.35

MINERALOGICAL COMPOSITION (X-RAY DIFFRACTION , OTHER)

Principal phase: Dioctahedrial smectite of montmorillonite type.

Impurities: 6-10% albite; 3-5% quartz; 2-3% micas; 1-2% calcite;
   trace of kaolinite.

Amorphous consituents: 2% silica; 0.4% aluminum oxide; less than
   0.2% iron oxide.

The major impurities can be easily eliminated by sedimentation.

CATION EXCHANGE CAPACITY

82.5 meq/100g; major exchange cations: Ca and Na

SURFACE AREA

N_2 area 46.0 m^2/g; fraction <2 μm : 86.1 m^2/g

MORPHOLOGY

Very fine particles of the order of 10 A thick, frequently showing
angles of 120°. The particles are formed in elementary domains with
a diameter of approximately 100 A and with fairly perfect mutual
orientation at about 60°.

THERMAL BEHAVIOR

DTA: endotherms at 691°C and 867°C, exotherm at 949°C.

Typical diagram of a montmorillonite. A weak endotherm at 560°C
is probably due to kaolinite. The water desorption endotherm is
at 150°C.

INFRARED SPECTROSCOPY

Typical spectrum of a montmorillonite with Fe^{3+} in octahedral
position. Traces of carbonate and quartz detectable.

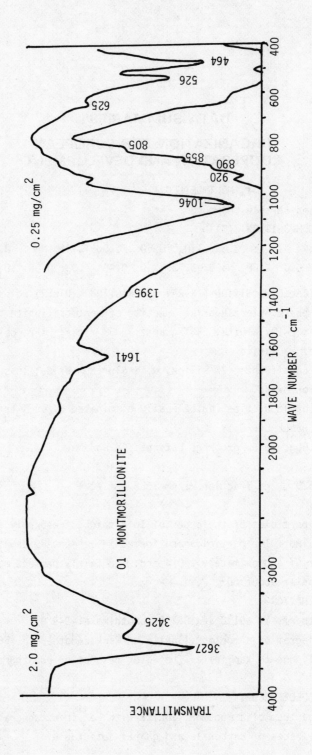

01 MONTMORILLONITE

                        02 LAPONITE

ORIGIN Synthetic

CHEMICAL COMPOSITION (%)

SiO_2: 66.03 Al_2O_3: 0.30 TiO_2: 0.02 Fe_2O_3: 0.06 MnO: 0.01

MgO: 29.03 CaO: 0.34 Na_2O: 3.19 K_2O: 0.04 P_2O_5: 0.02

Li_2O: 0.98

MINERALOGICAL COMPOSITION (X-RAY DIFFRACTION, OTHER)

Synthetic trioctahedrial smectite of hectorite type.

No impurities, except trace of carbonate (IR)

No amorphous consituents.

CATION EXCHANGE CAPACITY

73.3 meq/100g; major exchange cation: Na.

SURFACE AREA

N_2 area: 360 m^2/g

MORPHOLOGY

Particles with diameter between 200 and 300 A, and thickness less
than 20 A. The morphology is very different from that of natural
hectorite. The SAD patterns are composed of very diffuse bands,
characteristic of turbostratic smectites.

THERMAL BEHAVIOR

DTA: endotherm at 690°C, exotherm at 709°C, an exothermic-superposed-
on-endothermic system. The water desorption endotherm is at 130°C.

INFRARED SPECTROSCOPY

The spectrum is similar to that of natural hectorite.

A trace of carbonate is detectable.

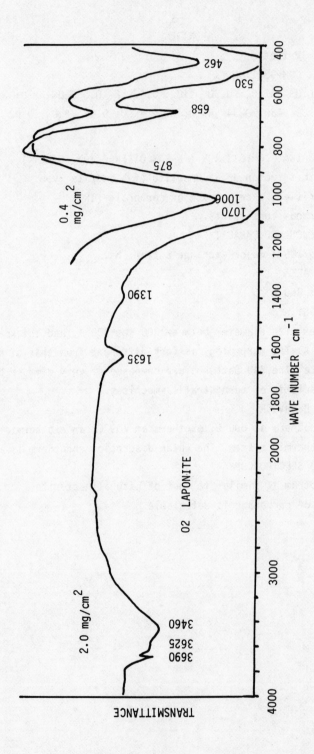

03 KAOLINITE (CHINA CLAY)

ORIGIN St.Austell, Cornwall, U.K.

CHEMICAL COMPOSITION (%)

SiO_2: 53.53 Al_2O_3: 43.95 TiO_2: 0.03 Fe_2O_3: 0.58 MnO: 0.01
MgO: 0.17 CaO: 0.2 Na_2O: 0.08 K_2O: 1.37 P_2O_5: 0.1

MINERALOGICAL COMPOSITION (X-RAY DIFFRACTION, OTHER)

Principal phase: well crystallized kaolinite (85-90%)

Impurities: 8-12% mica; 0.5-2% quartz; 2-3% feldspar;
  traces of tourmaline and a mixed layer mineral.
  Solution techniques show 11-13% mica; 0.5-1% quartz.
  IR shows trace of carbonate.
  The major impurities are easily eliminated by sedimentation,
  but muscovite mica has been reported to be still present in
  the <2 μm fraction.

CATION EXCHANGE CAPACITY

4.9 meq/100g.

SURFACE AREA

N_2 area: 6.6 m^2/g; fraction <10 μm: 8.8 m^2/g; fraction <2 μm:
  13.2 m^2/g.

MORPHOLOGY

Well formed pseudo-hexagonal particles, thin and of very variable
size. They are rich in dislocations and deformations, but each
particle is monocrystalline.

SAD: b-axis 8.96 A.

THERMAL BEHAVIOR

DTA: endotherm at 592°C, exotherm at 978°C. A small hygroscopic
moisture endotherm at 80°C.

TG: Total weight loss 12.65% (theory: 14%)

INFRARED SPECTROSCOPY

Typical spectrum of a well crystallized kaolinite. A trace of
carbonate is detectable.

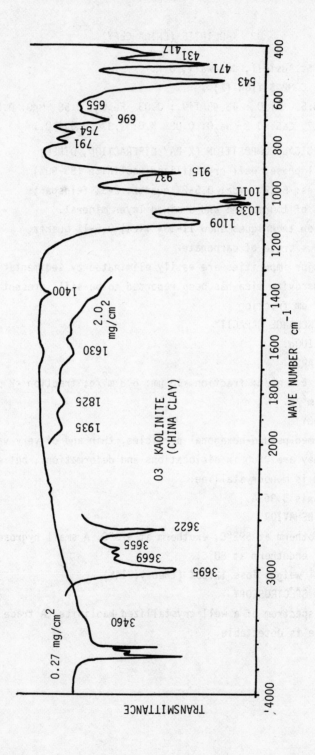

03 KAOLINITE
(CHINA CLAY)

04 ATTAPULGITE

ORIGIN Attapulgis, Florida, USA

CHEMICAL COMPOSITION (%)

SiO_2: 75.15 Al_2O_3: 9.72 TiO_2: 0.74 Fe_2O_3: 3.08 MnO: 0.10

MgO: 8.35 CaO: 2.03 Na_2O: 0.14 K_2O: 0.74 P_2O_5: 0.02

MINERALOGICAL COMPOSITION (X-RAY DIFFRACTION, OTHER)

Principal phase: attapulgite (palygorskite)

Impurities: 22-26% quartz; 4% dolomite and calcite;
    minor amounts of muscovite, feldspar, and chlorite (gibbsite?-DTA)
    The quartz is easily eliminated by sedimentation.

Amorphous constituents: 0.2% iron oxide, especially in the
    fraction <2 μm.

CATION EXCHANGE CAPACITY

9.0 meq/100g

SURFACE AREA

N_2 area: 50-83 m^2/g ; H_2O area: 172 m^2/g

MORPHOLOGY

Typical elongated crystals with lengths between 0.25 and 4 μm,
and a diameter of the order of 200 A. The needles are grouped in
bundles.

SAD: the axis of elongation is the c-axis with c = 5.2 A.

The micrographs show an impurity which is probably a lamellar
phyllite.

THERMAL BEHAVIOR

DTA: endotherms at 298°C and 495°C, exotherm at 917°C; The water
desorption endotherm is at about 160°C. Quartz is shown by an
endotherm at 573°C, and calcite by an endotherm at 740°C. The
doubling of the peak at about 300°C suggests the presence of
gibbsite or goethite.

TG: Total weight loss: 15.8% (theory: 19%)

INFRARED SPECTROSCOPY

Typical spectrum of a well crystallized attapulgite, however, the
spectral characteristics attributable to OH and SiO bonds are very
sensitive to the state of hydration of the sample.

Quartz and calcite are detectable.

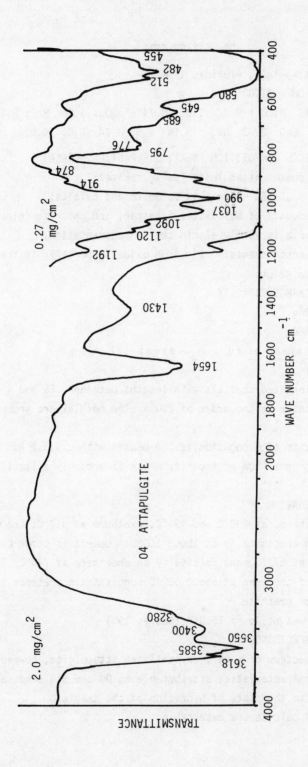

05 ILLITE

ORIGIN Le Puy en Velay, France

CHEMICAL COMPOSITION (%)

SiO_2: 54.58 Al_2O_3: 21.93 TiO_2: 0.90 Fe_2O_3: 7.41 MnO: 0.03
MgO: 3.76 CaO: 1.25 Na_2O: 0.39 K_2O: 8.88 P_2O_5: 0.48

MINERALOGICAL COMPOSITION (X-RAY DIFFRACTION, OTHER)

Principal phase: illite $2M_1$ (about 90%)

Impurities: 8% quartz; minor amounts of micas and feldspar;
  trace of carbonate (IR) and an unidentified impurity (IR).
  Major impurities are easily eliminated by sedimentation and
  the fraction <2 µm consists of practically pure illite.

Amorphous constituents: 2% silica;   1% aluminum oxide;
    0.1% iron oxides.

CATION EXCHANGE CAPACITY

26.6 meq/100g.

SURFACE AREA

N_2 area: 101 m^2/g.

MORPHOLOGY

The coarse fraction consists of micas and modified feldspars.
The fine fraction consists of small and irregular particles of
size <0.2 µm.

SAD: The b-axis measures 9.05 ± 0.05 A. The main impurity has a
b-axis of 9.2 ± 0.02 A.

THERMAL BEHAVIOR

DTA: endotherm at 562°C, exotherm at 875°C. The water desorption
endotherm is at 100°C and is unusually well developed.

INFRARED SPECTROSCOPY

Typical spectrum of a highly disordered illite containing octahedral
ferric ion. A trace of carbonate and an unidentified impurity are
detectable.

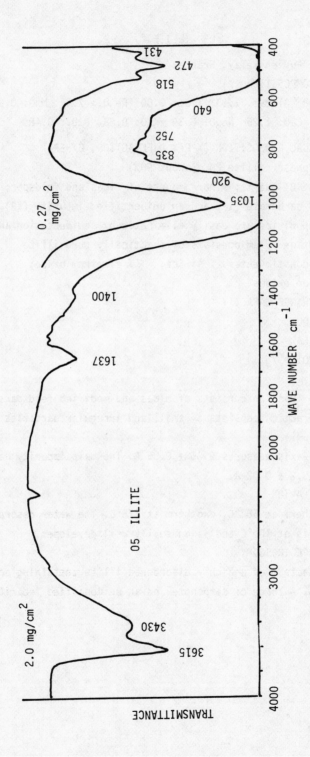

06   CHRYSOTILE

ORIGIN   Asbestos (Cassiar), Canada

CHEMICAL COMPOSITION (%)

SiO_2: 41.19-41.65 Al_2O_3: 0.52-0.83 TiO_2: 0.01-0.05

Fe_2O_3: 1.54-1.90   FeO: 0.13-0.25   MnO: 0.04   MgO: 41.5-43.2

CaO: 0.04-0.26 Na_2O: 0.02-0.07 K_2O: 0.01-0.03

P_2O_5: 0.08   Loss on firing: 13.45-13.94

MINERALOGICAL COMPOSITION (X-RAY DIFFRACTION, OTHER)

Principal phase: chrysotile   (97.7% - TG)

Impurities: 0.3% brucite (TG); 0.4% magnesite (TG);

  0.25% magnetite (TG); minor amount of a swelling clay,

  trace of carbonate (IR)

MORPHOLOGY

Tubular flexible fibers with a diameter between 200 and 500 A
grouped in bundles.

SAD: All fibers are elongated along the a-axis with a = 5.36 A.

THERMAL BEHAVIOR

DTA: endotherm at 699°C, exotherm at 809°C.

INFRARED SPECTROSCOPY

The spectrum is typical of chrysotile, but the reason for the
OH doublet in the stretching region, and the complexity of the
SiO stretching region are not understood.

A trace of carbonate is detectable.

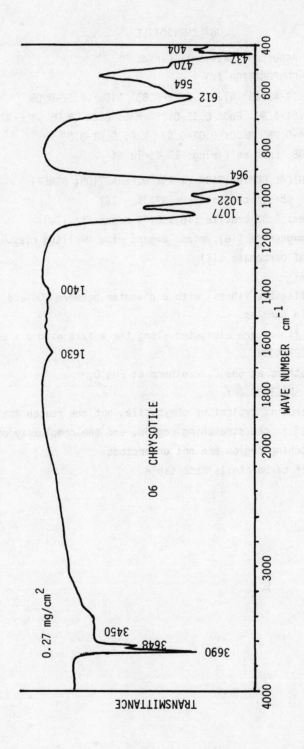

07 CROCIDOLITE

ORIGIN   Koegas Mine, Cape Province, South Africa.

CHEMICAL COMPOSITION (%)

SiO_2: 48.0-48.71 Al_2O_3: 0.2-1.02 TiO_2: 0.02-0.05

Fe_2O_3: 18.41-23.70 FeO: 18.65-21.59 MnO: 0.13-0.15

MgO: 0.67-2.32 CaO: 0.99-2.40 Na_2O: 5.16-6.00

K_2O: 0.07-0.1 P_2O_5: 0.01-0.04 Loss on firing: 1.12-3.38

MINERALOGICAL COMPOSITION (X-RAY DIFFRACTION, OTHER)

Principal phase: well crystallized crocidolite

Impurities: minor amount of calcite (IR)

MORPHOLOGY

Well formed laths.

THERMAL BEHAVIOR

DTA: exotherm at 434°C, endotherm at 899°C.; the exotherm
at 434°C is due to an oxidation reaction. A small endotherm
at 730°C suggests the presence of calcite.

INFRARED SPECTROSCOPY

Typical spectrum of crocidolite.

Carbonate is detectable.

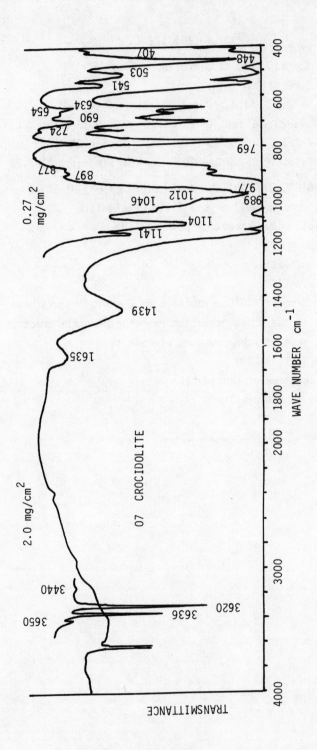

07 CROCIDOLITE

08 TALC

ORIGIN:Valle del Chisone, Piedmont, Italy

CHEMICAL COMPOSITION (%)

SiO_2: 60.36 Al_2O_3: 0.70 TiO_2: 0.07 Fe_2O_3: 0.95 MnO: 0.03

MgO: 31.41 CaO: 0.47 Na_2O: 0.04 K_2O: 0.03 P_2O_5: 0.15

MINERALOGICAL COMPOSITION (X-RAY DIFFRACTION, OTHER)

Principal phase: talc.

Impurities: 10% chlorite; minor amounts of calcite and quartz.
Chlorite is difficult to remove from the sample.

SURFACE AREA

N_2 area: 2.37 m^2/g; fraction <10 μm: 5.75 m^2/g

MORPHOLOGY

Fragments obtained by grinding are lamellar in shape with irregular
contours.

THERMAL BEHAVIOR

DTA: endotherm at 970°C; contaminating chlorite is probably
responsible for endotherms at 570°C, 600°C and 700°C, and
an exotherm at 870°C.

INFRARED SPECTROSCOPY

Typical spectrum of a magnesium talc with a small ferrous content.
Traces of carbonate are detectable.

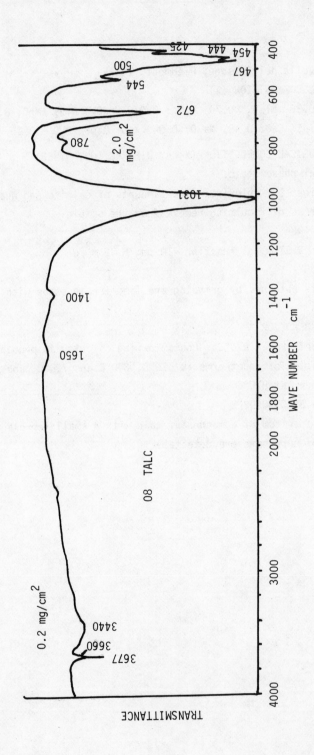

10 GIBBSITE

ORIGIN Synthetic material

CHEMICAL COMPOSITION (%)

SiO_2: 0-0.64 Al_2O_3: 65.2-65.45 TiO_2: 0 Fe_2O_3: 0.02-0.04
MgO: 0.01 CaO: 0 Na_2O: 0.36-0.38 K_2O: 0-0.03 Loss on
firing: 33.13-33.6

MINERALOGICAL COMPOSITION (X-RAY DIFFRACTION, OTHER)

Principal phase: gibbsite, but this synthetic material is
less well crystallized than some natural samples.

Impurities: minor

Amorphous constituents: silica: 0.1-0.3%; iron oxide: <0.05%

SURFACE AREA

N_2 area: 7.3 m^2/g.

MORPHOLOGY

Particles with hexagonal outline.

SAD: c = 9.7 A.

THERMAL BEHAVIOR

DTA: endotherms at 247°C, 327°C, and 533°C. The endotherms at 247°C
and at 533°C are due respectively to the formation and to the
dehydroxylation of boehmite, resulting from the thermal treatment.

TG: total weight loss: 34.15% (theory: 34.62%)

INFRARED SPECTROSCOPY

Typical of pure gibbsite.

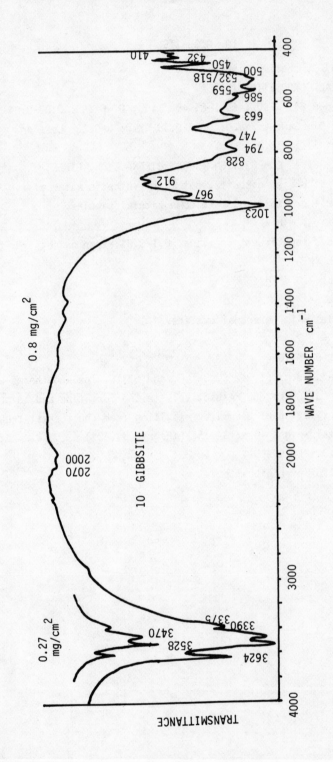

11 MAGNESITE

ORIGIN Austria

CHEMICAL COMPOSITION (%)

SiO_2: 5.43 Al_2O_3: 0.98 TiO_2: 0.06 Fe_2O_3: 1.44 MnO: 0.05
MgO: 42.50 CaO: 2.61 Na_2O: 0.13 K_2O: 0.09

MINERALOGICAL COMPOSITION (X-RAY DIFFRACTION, OTHER)

Principal phase: magnesite

Impurities: Very impure sample, contaminants are quartz, calcite,
  dolomite, mica, talc, and a chlorite.

SURFACE AREA

0.094 m^2/g, from permeability

THERMAL BEHAVIOR

DTA: endotherm at 661°C. A double endotherm at 770°C is due to dolomite,
and small quantities of quartz are shown by a weak endotherm at 573°C.

TG: total weight loss 46.3% (theory: 52.18%)

INFRARED SPECTROSCOPY AND RAMAN

The spectrum of magnesite with absorption bands due to talc, dolomite,
quartz, and muscovite.

The Raman spectrum has a single line at 1095 cm^{-1}(ν_1 for CO_3^{2-})

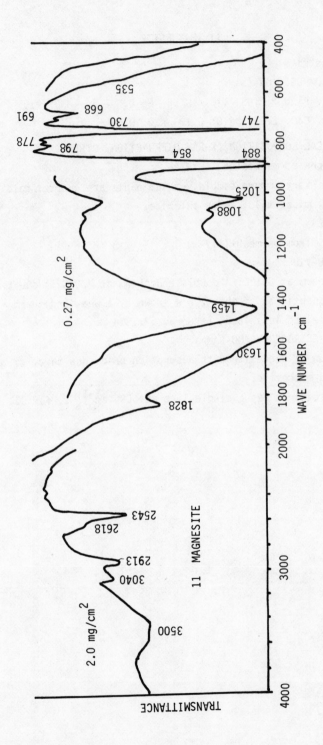

12 CALCITE

ORIGIN   Ireland

CHEMICAL COMPOSITION (%)

SiO_2: 0-0.18 Al_2O_3: 0.01-0.06 TiO_2: 0 Fe_2O_3: 0.05-0.09
MgO: 0-0.8 CaO: 54.00-55.58 Na_2O: 0.03-0.94 K_2O: 0-0.08
P_2O_5: 0.01 Loss on heating: 43.39-45.4 (CO_2: 43.6; SO_3: 0.08)

MINERALOGICAL COMPOSITION (X-RAY DIFFRACTION, OTHER)

Principal phase: well crystallized calcite

Impurities: minor amounts of muscovite, a chlorite, quartz,
   pyrite; IR: a silicate

SURFACE AREA

0.067 m^2/g from permeability

THERMAL BEHAVIOR

DTA: endptherm at 915°C

TG: weight loss 44.20% (theory: 43.9%)

INFRARED SPECTROSCOPY AND RAMAN

Typical IR spectrum of rather pure sample, however a very minor
amount of silicate gives a weak IR band between 1000-1100 cm^{-1}.

Raman: the Raman spectrum shows three lines:

713 cm^{-1} (ν_4 CO_3^{2-}); 1088 cm^{-1} (ν_1 CO_3^{2-}); 1439 cm^{-1} (ν_3 CO_3^{2-})

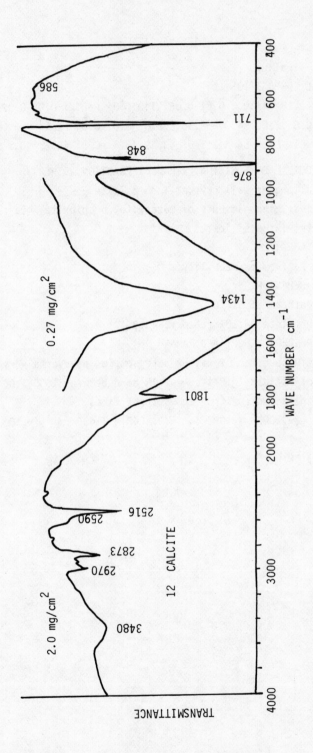

12 CALCITE

13 GYPSUM

ORIGIN Quarry of Vaujours, France

CHEMICAL COMPOSITION (%)

SiO_2: 0.94-1.27 Al_2O_3: 0.14-0.20 TiO_2: 0.01-0.03 Fe_2O_3: 0.07-0.10
MgO: 0.23-1.8 CaO: 32.0-33.6 Na_2O: 0.09-0.4 K_2O: 0-0.08 Cl: 0.17
CO_2: 1.25-2.25 SO_3: 42.3-43.40 Loss on heating: 20.65-22.17

MINERALOGICAL COMPOSITION (X-RAY DIFFRACTION, OTHER)

Principal phase: gypsum

Impurities: calcite, dolomite, quartz, trace of $SrSO_4$; IR: a silicate.

SURFACE AREA

0.075 m^2/g from permeability

THERMAL BEHAVIOR

DTA: endotherms at 171°C and 203°C, exotherm at 374°C. The endotherms
at 171 and 203°C are respectively due to the formation of the
hemihydrate qnd the anhydride. Calcite (about 7%) gives a double
peak at about 800°C.

TG: total weight loss: 19% (theory: 20.35%)

INFRARED SPECTROSCOPY AND RAMAN

Typical spectrum. Calcite and a silicate are detectable.

Raman: 1010 cm^{-1}, a S--O symmetric vibration band. (ν_1 SO_4^{2-})

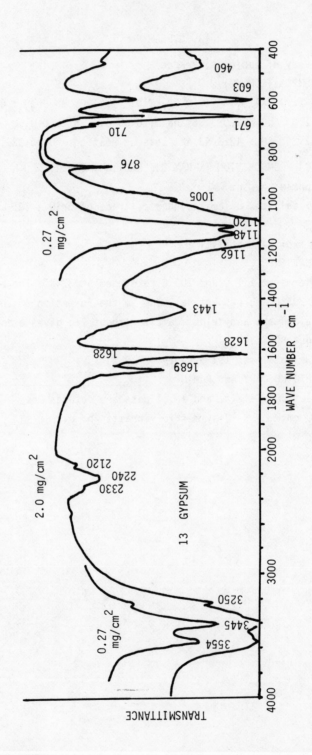

13 GYPSUM

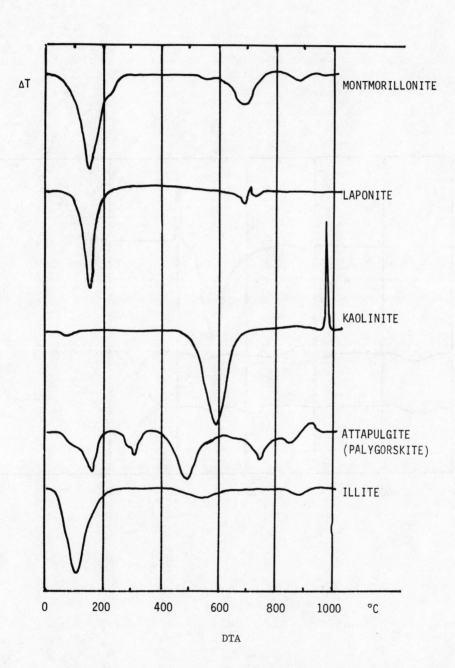

DTA

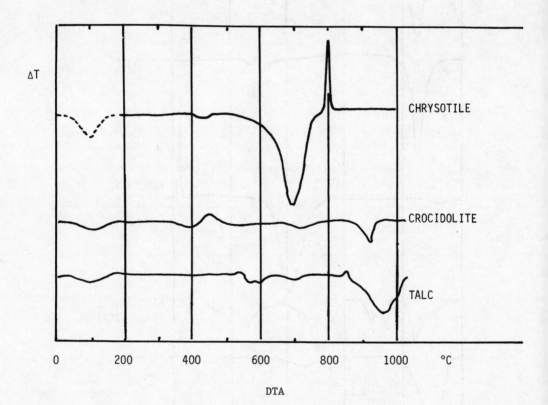

DTA

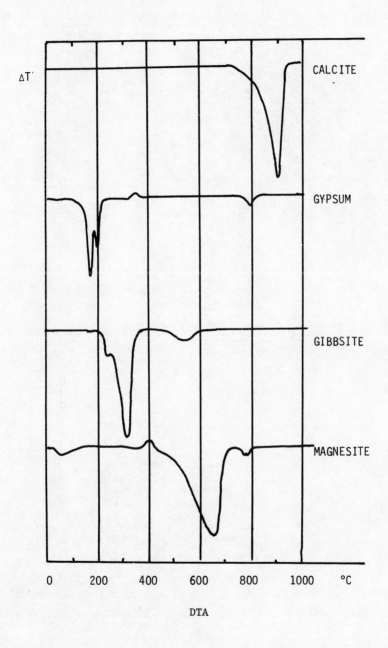

DTA

PART II

ORIGIN OF CMS SAMPLES

W. F. Moll

INTRODUCTION

Although clays bear mineral names, they are not well-definable
minerals in the sense of even carbonates or feldspars. In geological
deposits, the nature of the crystallites can change radically within
a few centimeters, and further separation from the raw sample can
change minerals considerably. The variations in properties, and
uncertainties introduced in separation cannot be overemphasized.
For example, the smectite in Wyoming bentonite , often assumed
to be a rather constant material, can show great variations. These
variations are a function of the particular stratum, amount of
overburden, extent of weathering, and other factors.

The name "Source Clays" for the suite of samples made available
through the Clay Minerals Society was chosen as not to imply that
this is a group of "standard" minerals to which any other mineral
of a particular group was to be compared. The Source Clays provide
investigators with a group of typical clay materials which will be
increasingly better characterized as the body of data on these
samples will develop.

Collecting and processing

Because of the large amount required for each material, the
aid was enlisted of companies exploiting deposits commercially.
A professional person thoroughly conversant with the geology and
mineralogy of the mining area carefully supervised the removal of
the batches.

To have a large amount of homogeneous material,some processing is
unavoidable , but what might be lost in the homogenization process,

such as the fabric of the clay, is more than compensated by the
advantage of having a large number of units of the same sample which
are reasonably identical.

The processing was held to an absolute minimum. The drying,
accomplished on steam-fed tray driers, at no time was done at a
temperature greater than 100°C. After drying the material was
heaped into a pile, and by quartering techniques was fed into a
roller crushing ("Raymond") mill for pulverization. The crushed
material was then put in large polyethylene bags in paperboard
drums.

Description of localities

The following descriptions of the origin of the samples
endeavor to give concisely the geological framework, theories
of paragenesis, and the precise location site. The discussions
and bibliographies are not intended to be exhaustive. They
are designed to present current thinking and to provide an
introduction to the literature.

KAOLINITE,GEORGIA, well crystallized (KGa-1)
KAOLINITE,GEORGIA, poorly crystallized (KGa-2)

Introduction

Along the contact between the Coastal Plain and Piedmont
in Georgia, kaolin deposits of unusual purity occur. The deposits
in central and east central Georgia are particularly desirable.
The sedimentary strata of the area are exceedingly complex. Only
now is the geologic history becoming at all clear. This kaolin
finds great use in paper coating, with lesser amounts used as
paint pigments, ceramics and fillers.

Some of the kaolin exhibits relatively sharp X-ray
diffraction patterns, whereas in others, the diffraction peaks
are broadened. These are termed, "high crystallinity" and
"low crystallinity", respectively. Samples of each were taken.
Both will be discussed in this section.

Geologic Framework

The sedimentary rocks of the Coastal Plain lap upon the
crystalline rocks of the Piedmont to the north in a line that
strikes northwest across Georgia (Figure 1). West of Macon, the
formations are mostly marine and have features amenable to mapping
(Zapp 1965). East of Macon, gradation into non-marine sediments
commences and distinguishing differences disappear in large sections
of the geologic column. It is in this complex sedimentary sequence
that the kaolin under consideration appears. For many years this
sequence was lumped with the Tuscaloosa formation and was considered
wholly Cretaceous.

Slowly, at first, some evidence began to provide a
depth of time for the units. The recent application of paly-
nology has provided the long-wanted breakthrough, though many
years will elapse before the sedimentary history is well understood. Some of the kaolin is Cretaceous, but much is lower Tertiary. Tschudy and Patterson (1975) have outlined the results of
the new approach and have given references to the investigators
responsible. Zapp (1965) mapped the sedimentary units in western

Georgia, where they can still be distinguished. Smith (1977), Chief
Geologist of the Georgia Kaolin Company, and other investigators,
have prepared a very preliminary correlation chart for central and
east Georgia (Figure 2).

The definitely Cretaceous clays tend to be better
crystallized than the Tertiary clays. Vermiform stacks of
kaolinite often are present. The deposits are not continuous.
Whether this implies deposition in lagoons, or post depositional
erosion of a sheet-like deposit is not known, although both may
be likely. Lignite is associated with the kaolin in some areas.

The Tertiary clays are finer in size and exhibit poorer
crystallinity. These Tertiary clay bodies seem to have a wider
areal extent than the Cretaceous ones. These clays often have a
greenish cast when fresh, but turn pinkish after exposure. The
vermiform stacks are rare. Again, lignite also occurs in
association with some deposits.

The crystalline rocks, principally schist, gneiss,
quartzite, and igneous rocks, once were considered pre-Cambrian.
Most of these rocks are now thought to be Paleozoic, even late
Paleozoic, and some of the igneous bodies early and middle
Mesozoic. The climate today is moist and warm-temperate, and
the crystalline rocks have weathered very deeply.

The well crystallized sample is from an almost certainly
Cretaceous zone (Figure 3). About 12 meters of sand overlie the
kaolin. Just at the contact, which is sharp, ferruginous concretions, often hollow and containing an evil-smelling fluid, occur.
The kaolin bed is 7.2 meters thick and rests on another sand. Near
the center of the bed, a band about 1 meter thick contains considerabl
smectite.

The poorly crystallized sample is from a Tertiary deposit
(Figure 4). About 12 meters of a crossbedded brown sand overlie
the kaolin. The contact is sharp. The kaolin bed is 11 to 12 meters
thick, and the upper 1 meter has a red stain. The clay, in turn,
lies on sand about 15 meters thick. Below the sand are crystalline
rocks.

Most investigators think that the kaolin resulted from deep weathering of the crystalline rocks. At this time, the sea extended up to the Piedmont. The products of the weathering were deposited in coastal lagoons. The presence of lignite supports this theory. The Cretaceous vermiform stacks could not have survived much transportation, however, and may have formed virtually in place.

Some workers have proposed that the kaolin resulted from alteration of volcanic ash. They feel the nature of the deposits is similar to those of ash deposits. However, this theory has not found general acceptance.

Paragenesis

Any general theory will have to explain the unusual purity (up to 90% kaolinite) of the deposits. Iron contamination is rather minimal, ranging in commercial deposits from 0.2 to 2 percent (expressed as Fe_2O_3). Titania, usually in the form of anatase, is present and ranges from 1 to 2 percent.

Obviously, much work remains to be done to explain this very valuable mineral resource. Through the efforts of the current investigators, a clearer picture will emerge. Patterson and Buie (1974) have described in some detail the geologic situation as it is now known.

Sampling

The well crystallized sample was taken under the direction of Mr. John M. Smith, Chief Geologist, Georgia Kaolin Company, on 4 October 1972. The collection site was the Coss Hodges Pit of the Georgia Kaolin Company, just north of Georgia Highway 24, about 17 kilometers west of Sandersville, Washington County at 32° 58'N, 82° 53'W, Transverse Mercator grid 3650500N, 315400E (See Figure 5). The Tabernacle, Georgia, 7.5' quadrangle map covers this area.

The pit was being actively mined and the sample was taken from a fresh face (Figure 6) in the zone above the band containing smectite.

The poorly crystallized sample was collected under the direction of Mr. John M. Smith on 4 October 1972. The collection site was the Purvis Pit of Georgia Kaolin Company, just west of Georgia Highway 17, about 14 kilometers north of Wrens, Georgia

at 33° 19'N, 82° 28'W (Figure 7). The Bowens Pond, 7.5' quadrangle map covers this area.

        The pit was being actively mined. The sample was taken
from a fresh face well below the red-stained band and in the
middle of the bed (Figure 8). Both samples were processed in the
specified manner. Both kaolin samples were kindly donated by
Georgia Kaolin Company.

BIBLIOGRAPHY

Huddlestun, P.F., W.E. Marsalis and S.M. Pickering, Jr. (1974)
        Tertiary Stratigraphy of the Central Georgia Coastal Plain,
        Georgia Geological Survey Guidebook Number 12.

Patterson, S.H. and B.F. Buie (1974) Field Conference on Kaolin
        and Fuller's Earth, November 14-16, 1971, Georgia Geological
        Survey Guidebook Number 14 (for the Society of Economic
        Geologists).

Smith, John M. (1977) Personal Communication.

Tschudy, Robert H. and Sam H. Patterson (1975) Palynological
        Evidence for Late Cretaceous Paleocene and Early and Middle
        Eocene Ages for the Strata in the Kaolin Belt, Central Georgia,
        Journal of Research, U.S. Geological Survey, p. 437-445.

Zapp, A.D. (1965) Bauxite Deposits of the Andersonville District,
        Georgia, U.S. Geological Survey Bulletin 1199-G.

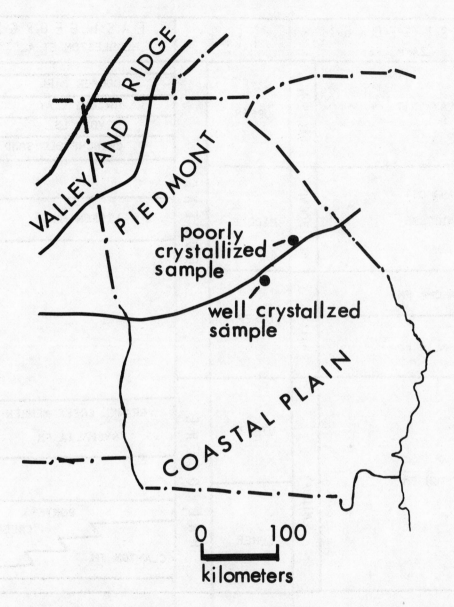

Figure 1. Map of Georgia showing location of kaolin samples.

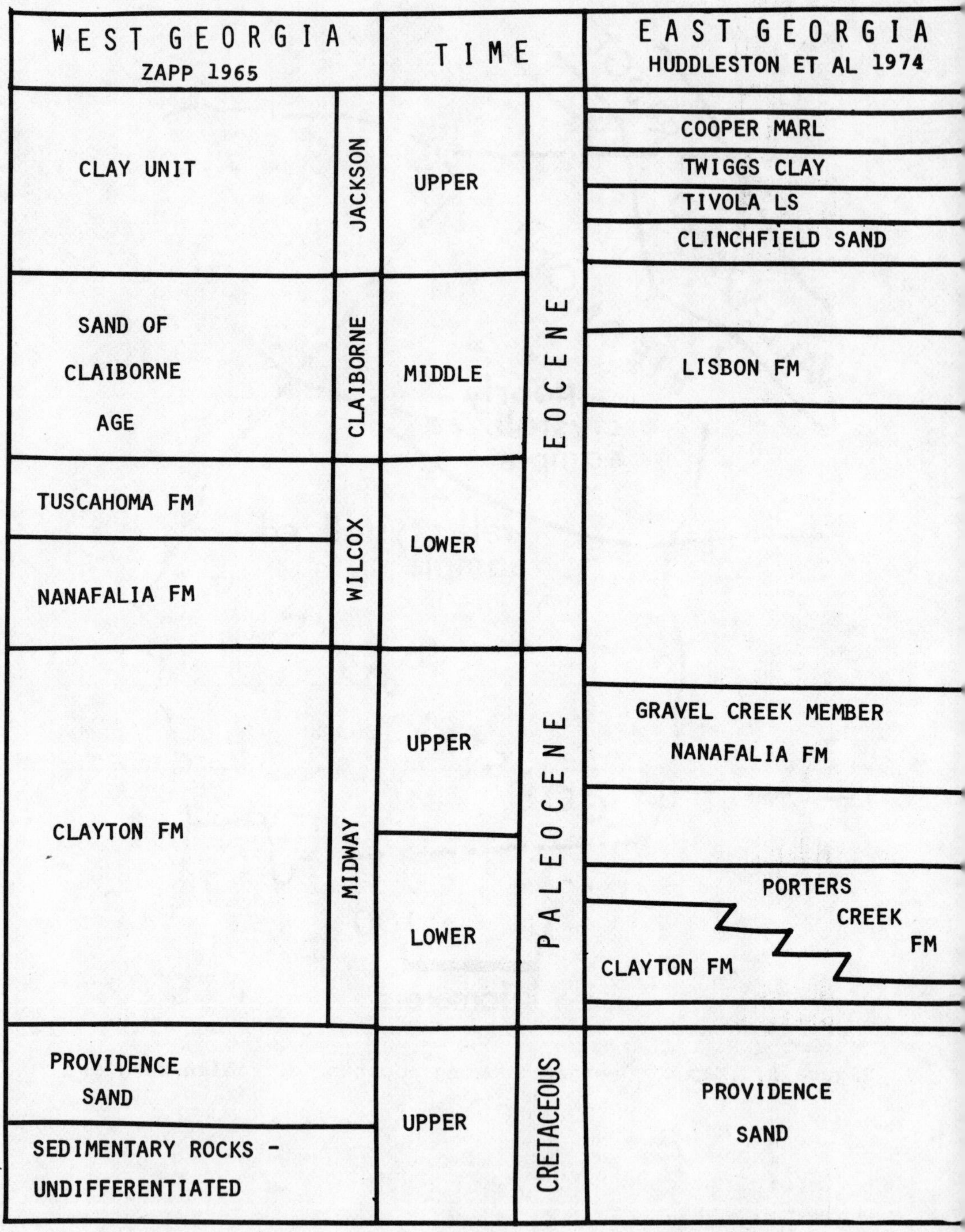

WEST GEORGIA ZAPP 1965		TIME		EAST GEORGIA HUDDLESTON ET AL 1974
CLAY UNIT	JACKSON	UPPER	EOCENE	COOPER MARL
				TWIGGS CLAY
				TIVOLA LS
				CLINCHFIELD SAND
SAND OF CLAIBORNE AGE	CLAIBORNE	MIDDLE		LISBON FM
TUSCAHOMA FM	WILCOX	LOWER		
NANAFALIA FM				
CLAYTON FM	MIDWAY	UPPER	PALEOCENE	GRAVEL CREEK MEMBER NANAFALIA FM
		LOWER		PORTERS CREEK FM CLAYTON FM
PROVIDENCE SAND		UPPER	CRETACEOUS	PROVIDENCE SAND
SEDIMENTARY ROCKS – UNDIFFERENTIATED				

Figure 2. Tentative correlation chart (Smith, 1977)

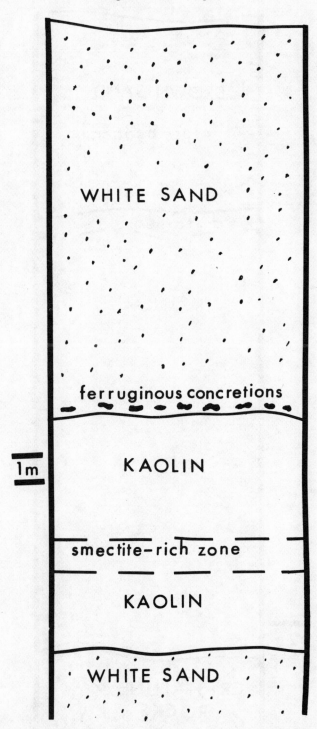

Figure 3. Lithology of Coss Hodges pit (well crystallized kaolin).

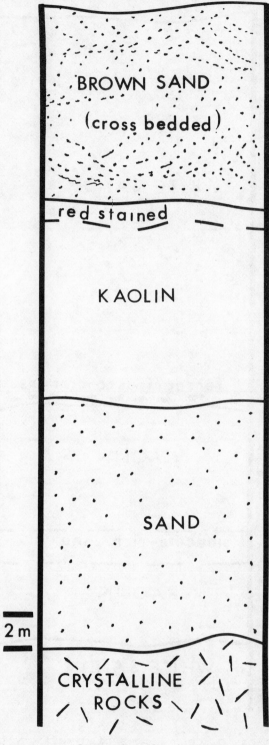

Figure 4.   Lithology of Purvis pit (poorly crystallized kaolin).

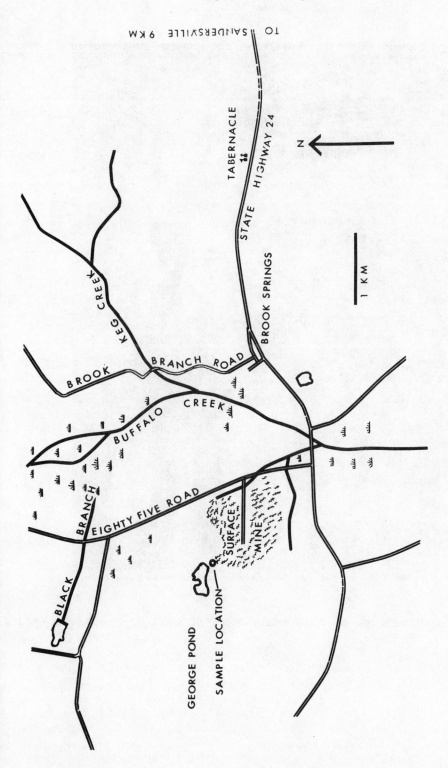

Figure 5.   Location of collection site of well crystallized kaolin sample.

Figure 6.   Photograph of collection site of well crystallized kaolin
sample.

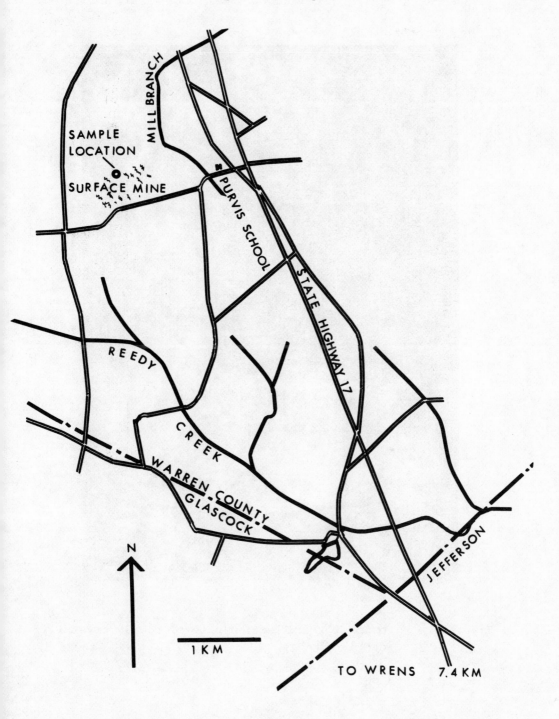

Figure 7. Location of collection site of poorly crystallized kaolin sample.

Figure 8.   Photograph of collection site of poorly crystallized
              kaolin sample.

MONTMORILLONITE, WYOMING (SWy-1)

Introduction

       The term, "Wyoming Bentonite", connotes a material consisting largely of sodium smectite and possessing outstanding binding ability and water-gelling capacity. Through these properties, Wyoming bentonite has found many industrial applications. Further, it has become a favorite for academic investigations even though it represents only a minor fraction of the world's bentonite reserves. Actually, bentonite occurs in a number of geologic units across Wyoming. The properties of bentonite in some of these units are vastly different from those usually associated with "Wyoming Bentonite". Often, processors blend material from various units to achieve certain desirable commercial properties. The unwary investigator may detect heterogenieties which may be entirely artificial. The sample here has the properties usually associated with the term, "Wyoming Bentonite". It was collected from a single site in the Newcastle formation.

Resume of Wyoming Bentonite

       Because of the confusion regarding the term, "Wyoming Bentonite", a brief resume of bentonite occurrences follows. Bentonite of various quality is widespread in the state. Osterwald, et. al. (1966) lists 18 counties with either prospects or operating mines. All commercial bentonites occur in rocks of the Cretaceous system, composed of shales, marls and argillaceous sandstones up to 1300 meters thick. Slaughter and Earley (1965) reviewed the deposits in the central and western portion of the state and Knechtel and Patterson (1962) described those in the Black Hills district in the northeast. This latter region is considered here in some detail. The sample was taken from this area. Also, this district has represented the major mining area for many years (Figure 1). Knechtel and Patterson recognized nine bentonite layers (Figure 2).

Bed A in the Newcastle formation, shows rapid thinning
and thickening, indicative of near-shore deposition.
The Mowry shale contains two beds. Bed B, ranging from
25 to 45 cm thick appears to contain divalent exchange
ions. The "Clay Spur" or "commercial" bed ranging from
0.7 to 2 meters thick occurs at the top of the formation.
This bed probably best typifies the properties generally
associated with Wyoming bentonite.

The Belle Fourche shale contains four beds. Bed D, near
the base of the formation, generally measures only 10
cm in thickness. This probably is the "Chalk Line"
described by Davis (1965). Bed E, known as the "Mud
Bed", is a dark impure bentonite measuring from 0.3 to
1 meter thick. Bed F is 0.6 to 2 meters thick. Its
informal name, "Gray-Red", describes the red staining
found in some areas. The numerous calcite veins ap-
pearing in the bed reduce interest in its exploitation.
Bed G, measuring 0.9 to 1.8 meters thick, occurs near
the top of the formation. Much of the bentonite contains
divalent exchange cations.

The next three formations, the Greenhorn formation, the
Carlisle shale, and the Niobrara formation contain little
or no bentonite, but the Pierre shale contains two beds.
Bed H, in the Gammon ferruginous member near the bottom
of the shale is a very impure bentonite. Although it is
only 0.5 meter thick, it persists over a rather wide area.
Bed I, however, ranges from 1 meter to 3.7 meters thick.
Shale partings occur within the bed in some places. The
preponderant exchange cation is calcium.

        All authors agree that these bentonites resulted from
volcanic ash falling into the sea. The source of the ash lay to
the west. By considering the grain size distributions of beds
in central Wyoming together with the aerodynamics of volcanic
explosion clouds, Slaughter and Earley (1965) felt they could map
and explain lobes of ash deposited by the clouds. The vulcanism
was thought to be associated with the emplacement of the Idaho
batholith.

Geologic Framework of the Newcastle Formation

The Newcastle formation varies over rather short distances both in thickness and lithologic character. Although it measures 18 meters thick in some areas, it thins in general to the east. The formation is composed of sandstone, at places cross-bedded, and interbedded with siltstone, sandy shale, carbonaceous shale, bentonite, and impure lignite. Terrestrial plant fragments occasionally appear as do worm burrows. All investigators, among them Knechtel and Patterson (1962) and Davis (1965), feel this indicates a near-shore environment. The volcanic ash fell near the beach and into lagoons. Concentration in lagoons and penecontemperaneous erosion produced the highly variable thickness of the bentonite.

Paragenesis

There can be no question that the bentonite resulted from a latitic volcanic ash that fell into the sea near the shore. Glass shards attest to the volcanic origin, and sedimentary features in the Newcastle denote the depositional environment.

Any proposed mechanism of alteration must explain loss of silica, retention of magnesium and the abundance of sodium ions. Slaughter and Earley (1965) addressed the first two problems, and felt that the marine environment could account for the exchange cation when the replacement series suggest it should have been supplanted by calcium or magnesium. They suggest that sodium rich solutions must have passed through the deposit at some time after alteration. The fact that the Wyoming bentonites are the only known large deposits containing appreciable sodium makes this question particularly interesting.

In the area of the sampling, the bentonite occurs in a lens 0.9 to 1.8 meters thick. The overburden, largely sandstones and shale amounts to approximately 2 meters (Figure 3). The Newcastle formation is almost flat-lying with a dip of 1° to the northwest. The area has a rather low relief. Owing to the semi-arid climate, most streams are intermittant.

Sampling

The sample was taken under the direction of Dr. Richards A. Rowland on 3 October 1972. The collection site was an area recently stripped of overburden on the Proctor Lease, approximately 50 kilometers west of Colony, in NE 1/4, SE 1/4, sec. 18, T.57N, R.65W. Crook County, Wyoming (see Figure 4). The Seeley, Wyoming, 15' quadrangle map covers this area.

The top 15 cm. of the exposed bentonite was stripped off, then the sample was removed (see Figure 5). It was taken over a meter above a brown zone. The sample was processed in the specified manner. This bentonite was kindly donated by Baroid Division, NL Industries.

BIBLIOGRAPHY

Davis, John C. (1965) Bentonite Deposits of the Clay Spur District, Crook and Weston Counties, Wyoming, Preliminary Report No. 4, Geological Survey of Wyoming.

Knechtel, Maxwell M. and Sam H. Patterson (1962) Bentonite Deposits of the Northern Black Hills District, Wyoming, Montana and South Dakota, Bulletin 1082-M, U.S. Geological Survey.

Osterwald, Frank W., Doris B. Osterwald, Joseph S. Long, Jr., and William H. Wilson (1966), Mineral Resources of Wyoming, Bulletin No. 50, Geological Survey of Wyoming (Revised by William H. Wilson).

Slaughter, M. and J. W. Earley (1965) Mineralogical and Geological Significance of the Mowry Bentonites, Wyoming, Special Paper 83, Geological Society of America.

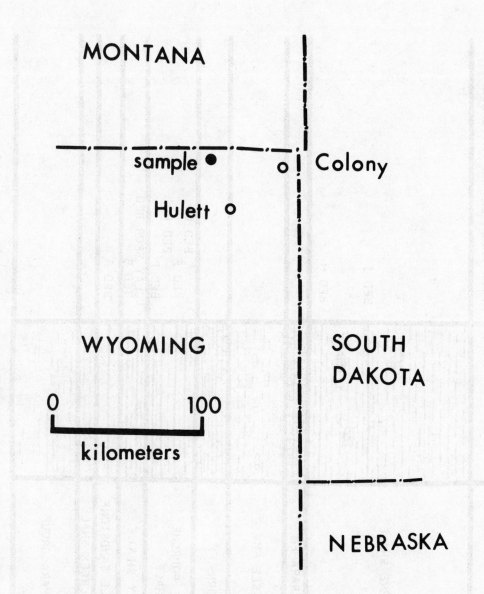

Figure 1.   Map of northeastern Wyoming showing location of Wyoming
            montmorillonite sample.

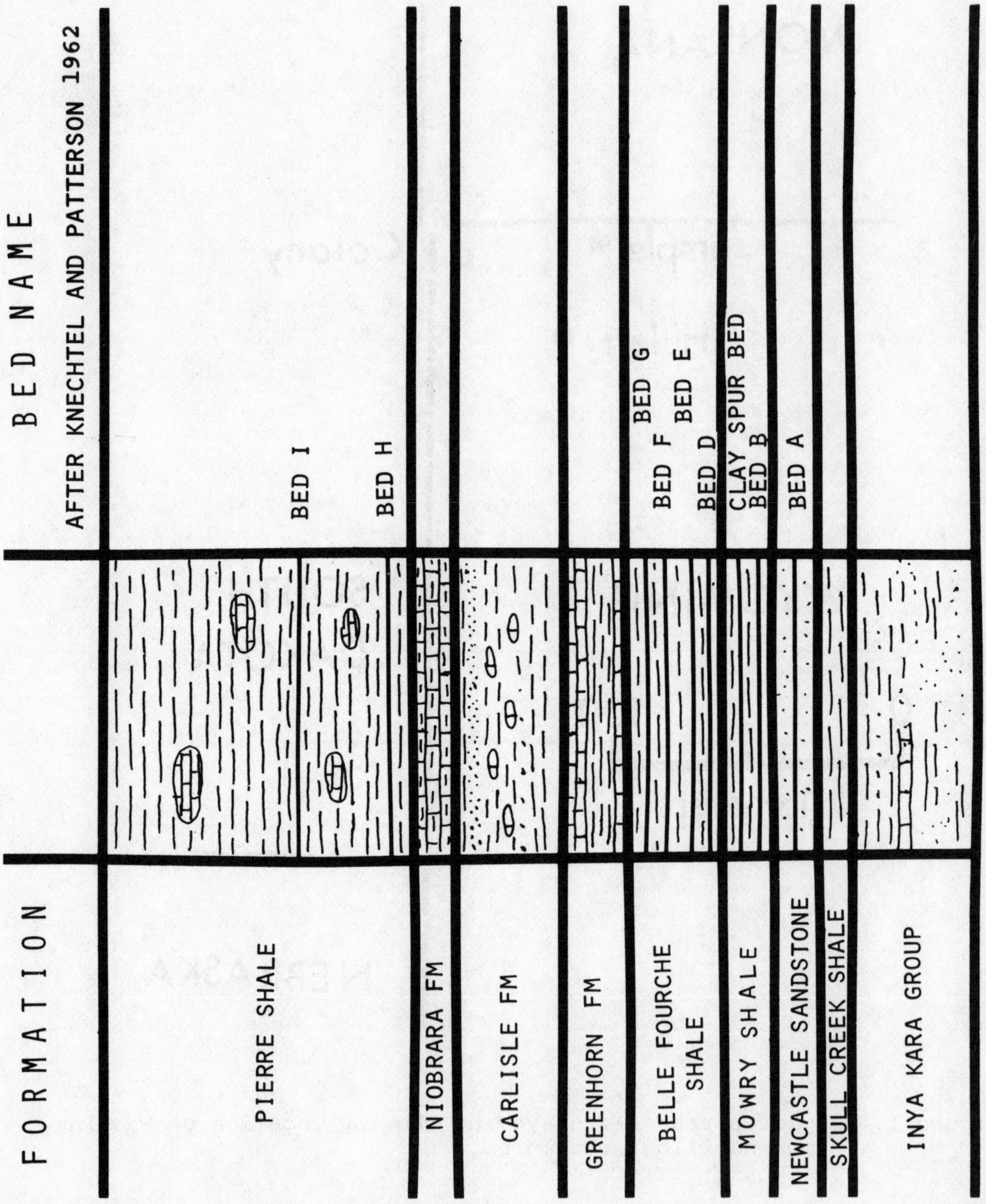

Figure 2. Bentonite beds in the Cretaceous of Wyoming.

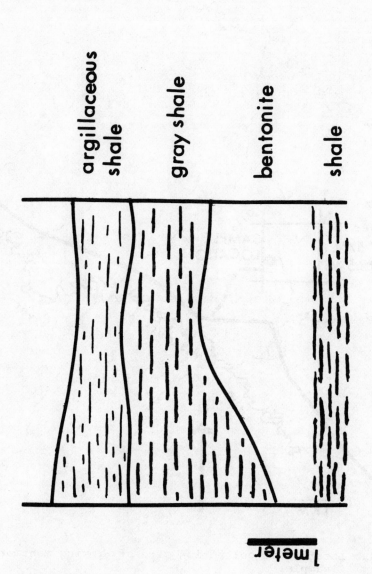

Figure 3. Lithology of the region of Wyoming montmorillonite sample.

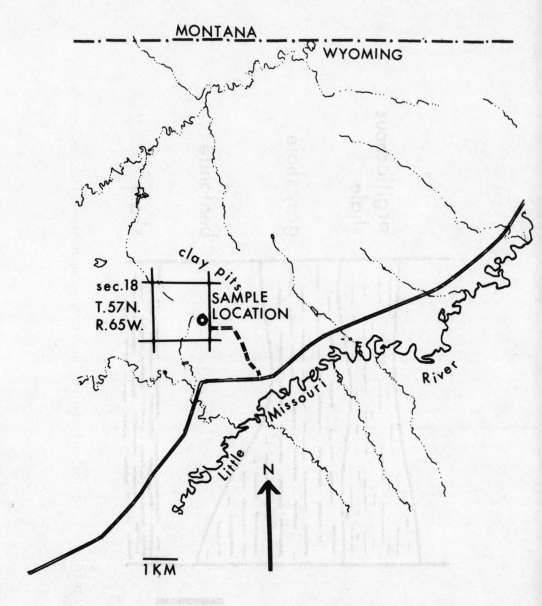

Figure 4.   Location of collection site of Wyoming montmorillonite
            sample.

Figure 5. Photograph of collection site of Wyoming montmorillonite sample.

MONTMORILLONITE, TEXAS (STx-1)

Introduction

     The Jackson group of east central Texas contains beds
of extremely white bentonite, a rather rare form of this substance.
In Gonzales County, the bentonite occurs in commercially mineable
amounts. It finds use in the tile industry as a binder owing to
its white-firing nature. Through processing it becomes a highly
desirable very white gellant for aqueous systems. Further, its
oil bleaching and catalytic properties have enjoyed considerable
use.

Geological Framework

     A broad band of sedimentary rocks, containing much
detrital volcanic material, today roughly parallels the Texas
Gulf coast about 130 kilometers inland. This sequence bears
evidence to explosive volcanism in Texas in mid-Tertiary time.
The abundant bentonitic materials present are well developed in
eastern Gonzales county, 100 kilometers east of San Antonio (Figure
1). Here an extremely white calcium bentonite occurs in the Manning
Formation of the upper Eocene Jackson group. Chen (1968) has examined
these white bentonites in very great detail.

     The area, gently undulating with a local relief of 80 meters
generally slopes to the southeast. The rainfall is somewhat over
60 cm. per year, but the Guadalupe River is the only perennial stream
flowing across the area. The regional dip is 1° to 2° to the southeast. Folding is absent here, but numerous faults of small displace-
ment have occurred.

     The areal lithology appears in the Geologic Column,
Figure 2. Nomenclature varies among the investigators (Eargle,
1959) with that of Chen used here.

     The particular bentonite comprising the sample occurs
at the very base of the Manning Formation and rests on a gray
friable sandstone either lowest Manning or uppermost Wellborn.
The bed is approximately 2 meters thick. As indicated in Figure 3,

it does not present a completely homogeneous appearance, but contains
several sections principally characterized by differences in fractures
and partings. These partings all are coated with a brown iron oxide.
At the very top, especially, the bentonite is excessively broken.
Directly above lies a cross-bedded sandstone containing pebbles of
white bentonite. Above the sandstone in turn lies approximately
0.5 meter of soil at the surface.

Paragenesis

        There is no disagreement that this white bentonite has
resulted from alteration of volcanic ash of rhyolitic composition.
The mechanism of conversion remains controversial, however. The
source of magnesium, which is much higher in the bentonite than in
locally-occurring unaltered ash, constitutes the main problem.
Chen believes that very fine ash fell into restricted coastal
bodies of water such as lagoons or abandoned delta distributaries.
He feels that seawater in these bodies donated the necessary magnesium. Jonas (1975), however, thinks that conversion did not take
place until after the ash beds were no longer directly in contact
with the marine environment. He has noted, from aerial photo-
graphs, that the ash beds tend to be altered where faults intersect
them. He theorized that subterranean brines rose up the fault
planes and infused the ash. The rich quantities of alkaline
earths in the brines aided the alteration of the ash. After
alteration the bentonites were subjected to some erosion as is
evidenced by bentonite pebbles in the overlying sandstone.

Sampling

        The sample was taken under the direction of Dr. Edward
Jonas, University of Texas, on 17 October 1972. The collection
site was the Helms pit of Southern Clay Products Company, just
south of U.S. Highway 90A about 10 kilometers east of Gonzales, at
29° 30' N., 97° 22' W. (See Figure 4). The Hamon, Texas 7.5'
quadrangle map covers this area.

        Active mining in the pit at the time exposed a fresh
face (Figure 5). The sample was taken from all the face below the
zone with numerous fractures. The iron stains on the fracture

surfaces were very carefully removed manually. The sample was
processed in the specified manner. This bentonite was kindly
donated by Southern Clay Products Company.

BIBLIOGRAPHY

Chen, Pei-Yuan (1968), Geology and Mineralogy of the White Bentonite
     Beds of Gonzales County, Texas. Ph.D. Thesis, University
     of Texas, Austin, Texas.

Eargle, D.H. (1959), Stratigraphy of the Jackson Group (Eocene)
     South Central Texas. Am. Assoc. Petrol. Geologists Bull.,
     43, p. 2623-2635.

Jonas, E.C. (1975), Personal Communication.

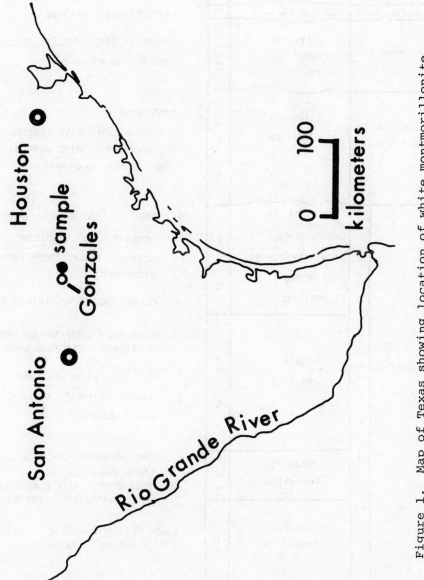

Figure 1. Map of Texas showing location of white montmorillonite sample.

AGE	FORMATION AND MEMBER	THICKNESS	LITHOLOGY
RECENT	ALLUVIUM		GRAVEL SAND MUD
PLEISTOCENE	TERRACE DEPOSIT	3	GRAVEL OF CHERT PEBBLES
MIOCENE	OAKVILLE FORMATION	30 M	CALCAREOUS SANDSTONE, SILTSTONE CONGLOMERATE AT BASE
OLIGOCENE OR LOWER MIOCENE	CATAHOULA FORMATION	35 - 120 M	TUFFACEOUS BENTONITE, VERY FINE SANDSTONE AND MUDSTONE, SHALE, BENTONITE, CONGLOMERATE NEAR BASE
UPPER EOCENE (JACKSON GROUP)	WHITSETT FM DUBOSE MEMBER STONES SWITCH SS MAIN MEMBER DILWORTH SS	45 - 60 M	4. MUDSTONE, INTERBEDDED SANDSTONE 3. INDURATED FINE SANDSTONE 2. MUDSTONE, SHALE, TUFFACEOUS SANDSTONE 1. FLAGGY SANDSTONE, INTERBEDDED SHALE
	MANNING FORMATION	45 - 90 M	2. MUDSTONE, SANDY SHALE, BENTONITE, SANDSTONE, TUFFACEOUS SANDSTONE 1. VERY FINE SANDSTONE, MUDSTONE, LIGNITE STRINGERS, ASHY BEDS, WHITE BENTONITE
	WELLBORN FORMATION	18 25 M	2. FINE INDURATED SANDSTONE, SANDY SHALE 1. CROSS BEDDED VITRIC ASHY SANDSTONE, SHALE, BENTONITE, FINE SANDSTONE
	CADDELL FORMATION	45 M	SANDY CLAY OR MUDSTONE SOFT SANDSTONE, BENTONITE
	YEGUA FORMATION		

Figure 2. Geologic column of Gonzales county, Texas (after Chen, 1968

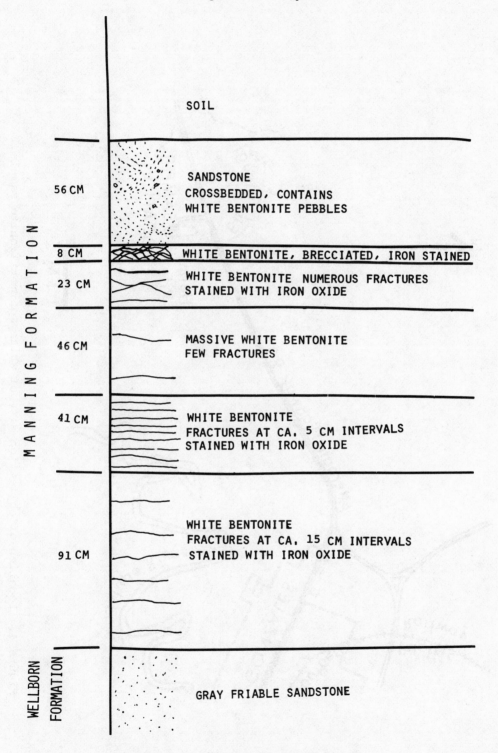

Figure 3.   Lithology of Helms pit.

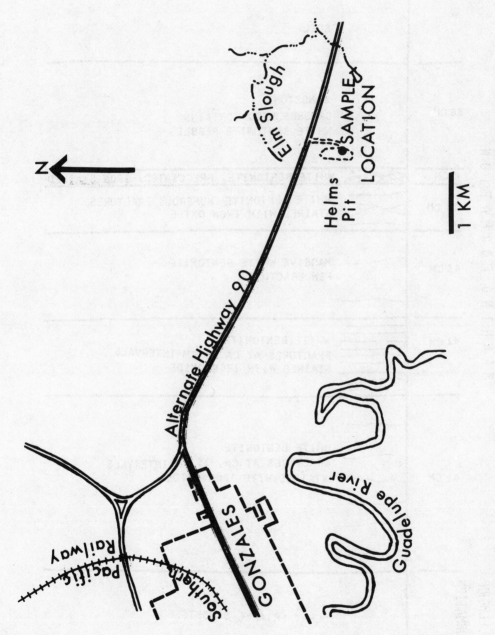

Figure 4.   Location of collection site of white montmorillonite sample.

Figure 5.   Photograph of collection site of white montmorillonite
            sample.

MONTMORILLONITE,ARIZONA (Cheto) (SAz-1)

Introduction

Occurring in the non-marine Bidahochi formation of
Pliocene age in northeastern Arizona are deposits of calcium
bentonite. Upon acid treatment this bentonite exhibits
unusually good bleaching and catalytic properties. Production
began in the Chambers district northwest of Sanders in the
1920's, and continues today in the Cheto district southeast
of Sanders.

Geologic Framework

The Bidahochi formation possesses a rather complex
depositional history. The portion containing the bleaching
clays, on the southwest flank of the Defiance Plateau, lies
unconformably upon Triassic (principally Chinle) and Permian
(DeChelly) formations (Figure 1). The lower member is composed
largely of mudstones and argillaceous, fine-grained sandstone
with some ash beds. The medial member, existing only outside
the area in question, consists of lava flows and detritus. The
upper member, which makes up sixty percent of the formation, is
largely poorly-cemented medium to fine-grained, argillaceous
sandstone. At the top, the member coarsens to contain lenses of
conglomerate, gravel and coarse sandstone. Also at the top, are
beds of calcareous sandstone, which can form resistant ledges.
At the bottom of the upper member are the bentonite beds.

Kiersch and Keller (1955), in their excellent paper,
suggest the following sequence of events. At the close of lower
Bidahochi time, streams cut into the landscape, creating numerous
channels and ponds. Then, perhaps contemperaneously with the
medial Bidahochi vulcanism to the west, a vitric ash fall of
quartz-latitic composition blanketed the area, clogging the
streams and choking the ponds. A period of erosion followed,
stripping the ash from the landscape, except in the channels and
depressions. Some of the ash was redeposited at this time. Then
the upper Bidahochi was deposited, probably subaerially, on the

region. Post-Pliocene has eroded the area again. Some stream
channels cut entirely through the bentonite beds (Figure 2).

The bentonite bed at the sample location measured
at least 1 meter thick. It lay apparently on the lower Bidahochi
member. Resting on the bed was approximately 2.5 meters of tan
sandy silt. The grayish tan color of the moist clay became grayish
white upon drying when exposed.

The sample location is about 16 kilometers east south
east of the locality of API 23, described by Kerr (1949).

Paragenesis

Kiersch and Keller (1955) feel that conversion of the
vitric tuff to smectite occurred before deposition of the upper
member of the Bidahochi. Chemically, this conversion principally
proceeded through gain of water and magnesium, and loss of silica
and the alkalis. They suggest the magnesium came either from the
stream and lake waters or from surrounding rocks.

Sloane and Guilbert (1967) in their electron-optical
study of the Cheto bentonite found that argillization developed
spontaneously and pervasively throughout the vitric shards. The
shards differed in susceptibility to alteration, therefore a
range of conversion is present. Orientation of the smectite
crystallites is a reflection of structures in the original glass.
Relic structures are present in even the completely altered material.

Sampling

The sample was taken under the direction of Dr. Robert
Secor and Mr. Robert Schwenkmeyer, Filtrol Corporation. The
collection site was an open pit, 10 kilometers east south east
of Sanders, in the SE 1/4, NE 1/4, NE 1/4, sec. 26, T.21N.,
R.29E., Apache County, Arizona (Figure 3). The Tolapai Springs
7 1/2' quadrangle covers this area.

The pit was being actively mined. After stripping off
the overburden, the exposed clay was carefully removed (Figure 4).
The sample was processed only in the specified manner. This
bentonite was kindly donated by the Filtrol Corporation.

BIBLIOGRAPHY

Kerr, Paul F. (1949) Reference Clay Minerals, A.P.I. Project 49.

Kiersch, George A. and W.D. Keller (1955) Bleaching Clay Deposits,
    Sanders-Defiance Plateau District, Navajo County, Arizona,
    Economic Geology, 50, 469-494.

Sloane, Richard L. and John M. Guilbert (1967) Electron-Optical
    Study of Alteration to Smectite in the Cheto Clay Deposit,
    Clays and Clay Minerals, 15, 35-44.

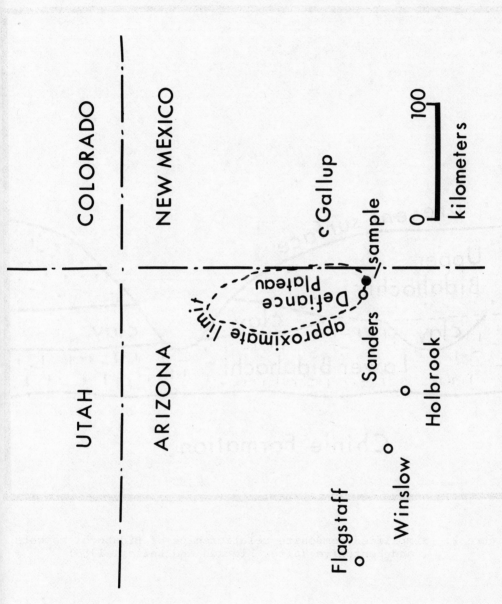

Figure 1.   Map of Four Corners area showing location of Cheto montmorillonite sample.

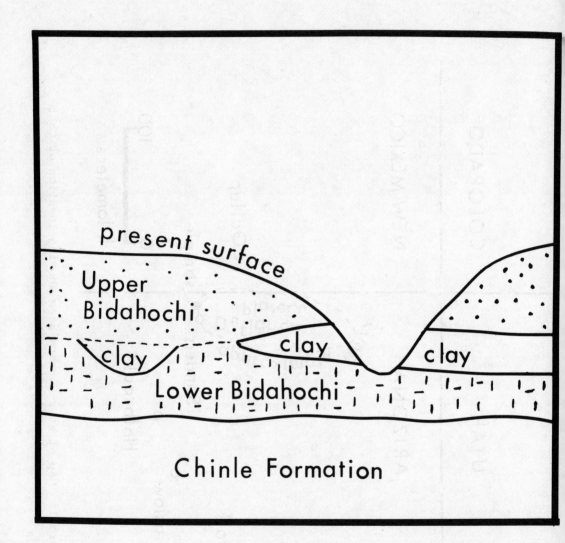

Figure 2.   Simplified composite relationships of Bidahochi members
            and bentonite (after Kiersch and Keller, 1955)

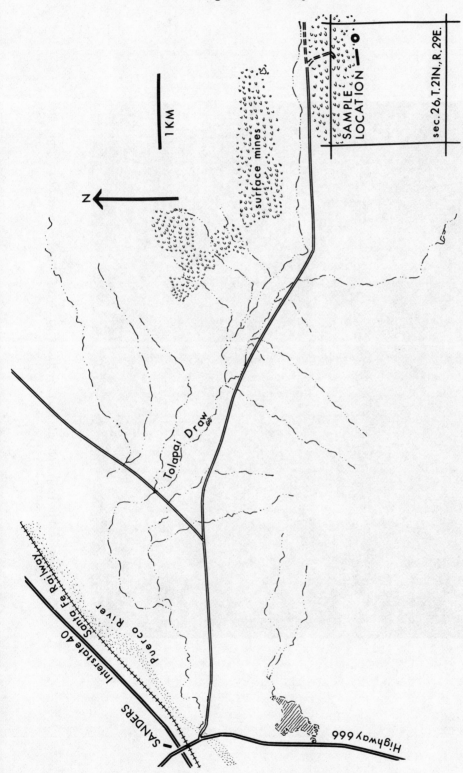

Figure 3. Location of collection site of Cheto montmorillonite sample.

Figure 4.   Photograph of collection site of Cheto montmorillonite
            sample.

SYNTHETIC MICA-MONTMORILLONITE (Syn-1)
(Barasym SSM-100)®

Introduction

Synthetic clay minerals remain fascinating, for they
open the possibilities to countless new compositions. The promise
of improving on nature and creating clays with unusual chemical
and physical properties has provided an interesting field of
research. Although many synthetic clays have been made in the
laboratory, few have enjoyed commercial manufacture. Scaling up
production from a few grams to many tons presents formidable
obstacles. These challenges have been overcome particularly with
trioctahedral smectites. Dioctahedral minerals have proved much
more difficult to synthesize on a large scale; Barasym SSM-100 is
one of these. The synthetic clays have enjoyed success particularly
as catalysts and viscosity control agents.

Synthesis

Precise techniques of synthesis remain proprietary, of
course. However, the patent by Granquist (1966) gives an excellent
generalized overview. In a typical example, an aqueous solution of
sodium silicate is passed through a hydrogen ion exchange column to
produce a silicic acid sol. Aluminum chloride hexahydrate then is
stirred into this sol. Coprecipitation of silica and alumina gel
results. This gel is filtered, washed and redispersed in water.
Now, sodium hydroxide is added. The mass is placed in an autoclave
provided with a stirrer and kept at 285°C. and 1000 p.s.i. for 48
hours. After cooling, the material is washed, then redispersed in
ammoniacal water and filtered. The ammonium hydroxide treatment
is repeated. The final product can either be tray dried and pulverized or spray dried to microspheroids of approximately 35 micron
size.

In this synthesis, other sources of silica or alumina can be used. Further, other ions may be added to provide particular desirable properties. The resulting material has the general formula:

$$\left[(Al_4)^{oct.} \ (Si_{8-x}Al_x)^{tetra.} \ O_{20}(OH, F)_4 \right] \left[wNH_4^+ + y \ Al(OH)_{3-z}^{z+} \right]$$

Nature of the Synthetic Mineral

Barasym SSM-100 is a mixed layer mica beidellite. It is described in detail by Wright, Granquist and Kennedy (1972). Note that this particular variety has not been calcined. The material is an irregularly mixed layering of 10.4 A (A) and 12.5 A (B) units. The ratio of A to B is 2 to 1. The MacEwan transform indicates the A and B stacking is random, with perhaps a slight amount of AB ordering.

Electron optical studies show the material crystallizes into rather irregular platelets, approximately 1000 A across. The platelets average 50 A in thickness.

Manufacture

This sample was produced in the Wallisville (Texas) synthetic minerals plant of NL Industries in 1972. It was kindly donated by NL Industries.

Bibliography

Granquist, William T. (1966) Synthetic Silicate Minerals, U.S. Patent 3,252,757.

Wright, Alan C., William T. Granquist and James V. Kennedy (1972), Catalyis by Layer Lattice Silicates I. The Structure and Thermal Modification of a Synthetic Ammonium Dioctahedral Clay, Journal of Catalysis, 25, p. 65-80.

HECTORITE,CALIFORNIA (SHCa-1)

Introduction

The type locality of the extremely white, lithium-bearing trioctahedral smectite, "hectorite", lies in a complex igneous terrain in southeastern California. Its white color and its great ability to gel water has found it many uses particularly in paints and cosmetics. Hectorite also is an excellent clarifier of beverages.

Geological Framework

Hectorite occurs associated with volcanic rocks in the Mojave Desert near Hector, California (Figure 1). The region is underlain by the Red Mountain andesite series of lower Pliocene Age. Tuff and agglomerates of angular lava fragments embedded in tuff are interbedded in the lava flows. Over the Red Mountain series lies a bed of andesitic gravels of alluvial fan origin that interfinger with Pliocene and lower Pleistocene lake sediments. These gravels are, in turn, overlain by a series of brown lacustrine mudstones. These deposits are covered over much of the area by an olivene basalt flow of Pleistocene or Recent Age from nearby Mt. Pisgah and also by more recent alluvial fans and eolian sands. The actual nature of the hectorite deposit, and its mode of origin, remain controversial.

Foshag and Woodford (1936) thought the deposit is associated with a small anticline. In their interpretation (Figure 2a), the mine face shows a series of interbedded sands and clays dipping away from the anticline. Whereas the upper part of the series is probably eolian, the lower part is tuff-aceous. Bed 2 is brown, altered dacitic tuff. The tuff contains chalcedony and calcite concretions. The hectorite is a secondary mineral developed at the surface and in cracks in the tuff.

Kerr (1949) concurred that the deposit is associated with an anticline. Bentonite layers alternate with layers of

less altered ash, silty clay, and limestone beds (Figure 2b).
The tight folding in places has intensely sheared the hectorite
and thickened the thin layer into a thicker bed which is
commercially exploitable.

Ames, Sand and Goldich (1958) provide a much different
and more complex picture. Instead of an anticline, they propose
a northwest-trending travertine ridge. This ridge is associated
with faulting in the Barstow-Bristol trough. Lapping onto the
travertine ridge are progressively zeolitic ash, hectorite, brown
mud, zeolitic ash and brown mud. This series is overlain unconformably by the gravels and the olivine basalt flow. (Figure 2c).
The travertine, although containing pyroclastic material, contains
no dolomite. Hectorite does appear interbedded with the travertine.
The hectorite bed itself contains nodules of chalcedony and calcite.
The brown mud contains hectorite in its base and grades into
analcime tuff.

Paragenesis

Foshag and Woodford concluded that the hectorite resulted
from alteration of dacitic volcanic ash. They proposed no mechanism,
but Ross and Hendricks (1945) thought highly magnesium solutions
were necessary to generate the hectorite.

Kerr felt the hectorite formed from alteration of volcanic
ash or even a lava flow. The abundant calcite was derived by
recrystallization from the overlying limestone.

A more elaborate origin is postulated by Ames, Sand
and Goldich. They proposed an ancient lake astride a fault from
which hot springs periodically flow. The hot springs built the
travertine ridge. Ash clouds of dacite composition occasionally
blanketed the area. When the springs were inoperative, the ash
altered into analcime. If however, the springs flowed during
the ash fall, the ash altered into hectorite. The necessary
lithium and fluorine came from the spring water. However, because
the travertine contains no magnesium, this element came from the
lake water.

The Findings of Sweet (1977)

        Sweet, the current Hector Mine engineer, enjoys year-
around observation of the deposit and access to extensive drilling
records, advantages unavailable to earlier workers. He has kindly
prepared for this project an analysis of the current findings.
This valuable contribution to the understanding of the deposit
should be quoted verbatim:

    Hectorite is the dominant clay mineral in a tele-
    thermally-altered series of Cenozoic volcanic ash beds.
    These beds occur interbedded both with cherty limestone
    deposited by hot springs and with lake sediments in the
    Mojave Desert. The deposit parallels the northeast
    side of a major northwest-trending strike-slip fault
    zone (Figure 3).

    The vitric ash fell into a partially restricted and
    protected environment of linear nature. A shallow
    trough, filled by a lake, trended northwest. The
    lake's southwest shore was paralleled by a traver-
    tine ridge. The shoreline was a fault-terrace. The
    tuff beds in the upper series of lake sediments lay
    interbedded with mudstones and clay stones. The tuff
    has been altered to clay minerals or zeolites. Adjacent
    to the deposit, and to the northeast, significant amounts
    of colemanite exist at depth in an evaporite section
    consisting of rhythmic laminations of anhydrite, clay
    and calcite, and beds of claystone.

    During a period of post-depositional faulting and warping,
    the mudstones and claystones overlaying the deposit were
    partially eroded. This exposed the more resistant
    travertine limestone beds. Thus, the travertine beds
    became a ridge, higher than the alluvial gravels that
    overlaid the lake sediments. A basalt flow of recent
    origin then filled the trough and paralleled the
    travertine ridge for approximately 12 kilometers.
    This flow almost completely covers the hectorite
    deposit.

    The vitric tuffs had their probable source to the north-
    west in the Fort Cady Mountains, where a series of
    volcanic ashes, tuffs, obsidian (perlite), and andesites
    intertongues with lake sediments. Minor gypsum and
    travertine tufa occurs there. Interestingly, a zeolite
    tuff bed that outcrops in this area is traceable to the
    hectorite deposit, over 12 kilometers to the southeast.
    In the deposit it serves as a "marker bed". This tuff
    bed is altered to clinoptilolite, erionite and hectorite.
    The relative amounts of each depends on its proximity to
    the deposit.

    The lacustrine-pyroclastic facies represents a distinct
    stratigraphic unit. The deposition of the tuff was

contemporaneous with extensive hot spring activity that
provided hot, lithium-rich solutions. These solutions
altered the vitric tuffs deposited in the terraced pools
and on the fault-terrace shoreline paralleling the ridge
built by the hot springs. Circulating saline lake
waters probably provided the necessary magnesium.
Development of hectorite apparently was restricted to
the on-shore zone. Off shore, in the lake, a brown
mud containing a non-hectorite smectite developed. The
high boron content of the hectorite indicates deep-
seated leaching of colemanite from the older, buried
evaporite series by the ascending hydrothermal solutions.

Alteration of the tuffs is partial to complete. This
alteration apparently continued after burial of the
tuffs by successive claystones, travertine limestones,
and ashfalls, which are interbedded in the deposit.
Travertine commonly occurs at or near the base of the
commercial clays. Chert nodules are prevalent in the
more completely altered zones containing a high percentage of clay. To the northwest, the tuffs have been
altered to a high-magnesium aluminum montmorillonite.
To the southeast, the smectite has a higher lithium
content.

Hectorite seems to be unique in that it is a commercial
lithium smectite of telethermal origin. Extensive core
drilling is continuing in order to delineate the total
extent and nature of the smectite deposits. The drilling
also will provide a better understanding of the depositional
and structural controls resulting from minor post-depo-
sitional normal faulting and slump structures associated
with this faulting.

Sampling

Owing to the mining difficulties associated with the
intimate intermixing of hectorite and the calcite, the sample
could not be taken in the usual fashion. Instead, it was
collected from the plant stockpile. The stockpile is continually
replenished with newly-mined material. Mr. A. J. Higgins, NL
Industries, collected the sample in November 1972. The mine is
located about 5 kilometers south of Hector, California in N 1/2,
sec. 35, T.8N., R.5E. The Cady Mountains 15' quadrangle covers
this area (Figure 4).

The mine pit is shown in Figure 5. The sample was
processed in the specified manner. The hectorite was kindly
donated by Baroid Division, NL Industries.

BIBLIOGRAPHY

Ames, L.L., Jr., L.B. Sand and S.S. Goldich (1958), A Contribution
    on the Hector, California, Bentonite Deposit. Economic
    Geology, 53, p. 22-37.

Foshag, W.R. and A.O. Woodford (1936), Bentonite Magnesium Clay-
    Mineral from California. American Mineralogist, 21, p. 238-244.

Kerr, Paul F. (1949), Reference Clay Minerals, A.P.I. Project 49.

Madsen, Beth M. (1970), Core Logs of Three Test Holes in Cenozoic
    Lake Deposits Near Hector, California. U.S.G.S. Bulletin 1296.

Ross, Clarence S. and Sterling B. Hendricks (1945), Minerals of the
    Montmorillonite Group. U.S.G.S. Professional Paper 205-B.

Sweet, Wilbur E., Jr. (1977), Written Communication.

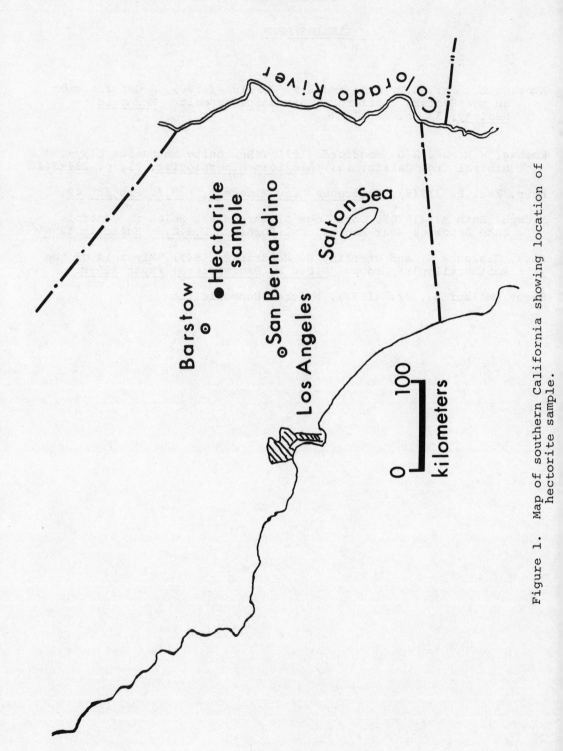

Figure 1. Map of southern California showing location of hectorite sample.

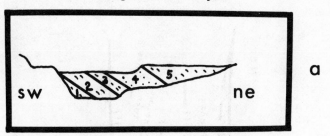

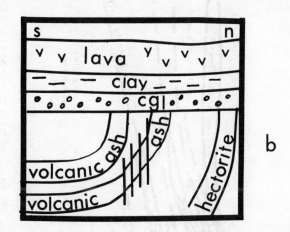

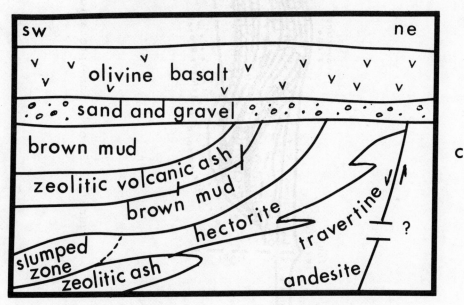

Figure 2.   Interpretations of the hectorite deposit:

   a. After Foshag and Woodford (1936)
   b. After Kerr (1949)
   c. After Ames, Sand and Goldich (1958)

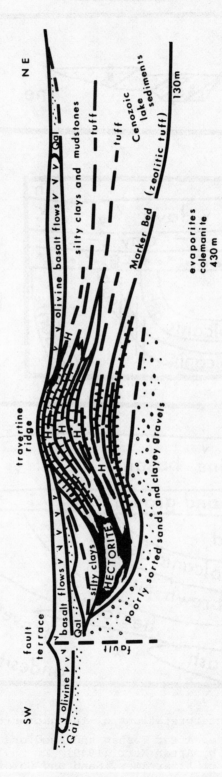

Figure 3. Sweet's interpretation of the hectorite deposit (1977).

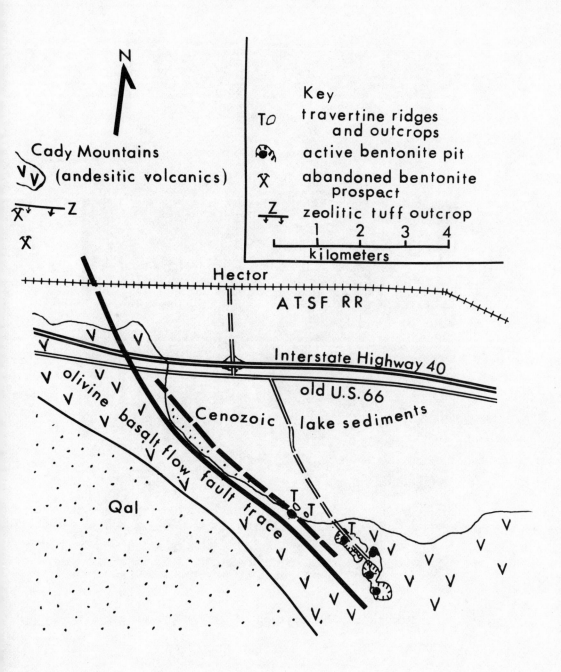

Figure 4.   Areal geology of hectorite deposit and location of pits.

Figure 5. Photograph of hectorite mine.

ATTAPULGITE,FLORIDA (PF1-1)

Introduction

The Hawthorn formation, a complex lithologic unit of
Miocene age and primarily of marine origin, contains several
deposits of clay minerals. The Meigs-Attapulgus-Quincy district
of Georgia and Florida is particularly rich in attapulgite*,
montmorillonite and sepiolite. Near the Georgia-Florida border,
rather pure attapulgite occurs in mineable amounts. Exploitation
began in 1895 and continues on a large scale today. Although oil
clarification, the original use of attapulgite, has decreased over
the years, the mineral continues to enjoy use as an insecticide
carrier, a thickener for salt-water drilling fluids, and adsorbent
granules.

> *"Palygorskite" has supplanted "Attapulgite" in nomen-
> clature. However, the latter term is used here
> because of its continued use in the mining area
> in question.

Geology

The district lies in the Coastal Plain, a region of low
dip and gentle relief (Figure 1). It is contained in the Gulf
Trough, a feature striking northeast from the Gulf of Mexico and
characterized by a thickening of the sediments. Although some
investigators describe the Trough as a graben, others now feel
it results from a strait that separated peninsular Florida from
the mainland in Oligocene and Miocene time. Because of rapid
facies changes, extensive solution of limestones and poor surface
exposures, the lithology of the area presents very serious problems
in interpretation. Excellent reviews by Patterson and Buie (1974)
and Patterson (1974) describe the situation in detail.

The Hawthorn formation measures at least 30 meters
thick in this area and lies apparently conformably atop the
Tampa limestone. In turn, it is overlain, perhaps unconformably,
by the Miccosukee formation. It consists of fine-to-medium
grained quartz sand, silt, sandy calcite and dolomite beds, some
phosphate material, opal, and clay. The attapulgite occurs as

large lenses which form a discontinuous bed. In the southern
portion of the Meigs-Attapulgus-Quincy district, the area in
question, the attapulgite lies well below the Miccosukee-
Hawthorn contact (Figure 2).

The sample was taken in the Luten Mine, approximately
1 kilometer south of La Camelia mine, measured by Patterson
(1974, p. 38). The overburden consists of approximately 8 meters
of sand which contains some montmorillonite. This is undoubtedly
in the Hawthorn formation. Directly beneath this sand are two
layers of attapulgite, each 0.7 meter thick separated by 0.7 meter
of limestone. Under this clay unit is 1.8 to 2.4 meters of sand
which, in turn, in underlain by 1.5 to 1.8 meters of attapulgite.
Only the topmost layer of the upper clay unit was included in
the sample.

Paragenesis

Because of their fine, fibrous nature, the attapulgite
particles almost certainly formed in place, for they could not
have withstood transportation. Shallow marine conditions represented the depositional environment. Evidence includes shallow
water diatom fauna, mudcracks, channel-fill deposits, algal heads,
and dolomite beds and numerous other indications. The actual
origin of the attapulgite remains controversial, however.
The source material is uncertain. Whether it formed penecon-
temporaneously with deposition, or from later diagenetic changes
is uncertain. The presence of kaolinite, montmorillonite and
sepiolite adds to the confusion. The reviews of Patterson and
Buie (1974) and Patterson (1974) discuss the problem and present
the conclusions of various workers.

Volcanic ash has been considered the ultimate source
material by some investigators. However, most other workers
consider the evidence to be much too indirect and scanty. They
feel that not only is the amount of identifiable volcanic material
much too small to indicate an ash fall, but also no relic structures
exist. The occasional reported shards are probably diatom fragments. Zeolites, so often associated with altered ash, are absent.

Any theory of origin must explain silica and alumina
concentrations far in excess of those in the open ocean. Some

workers feel that the silica and alumina were brought down by
streams into the restricted marine embayments. With magnesium
and alkalis provided by the seawater, all the needed constituents
for attapulgite were present. Thus, the attapulgite and sepiolite
crystallized contemporaneously with other materials typical of
such an environment,as dolomite.

Origin of the montmorillonite and kaolinite is insep-
arable with an understanding of the attapulgite deposits. Some
of the montmorillonite probably formed along with the attapulgite.
However, much of it may have formed during diagenesis and later
by deep weathering. Most of the kaolinite likely is a weathering
product of attapulgite-derived montmorillonite.

Sampling

The sample was taken under the direction of Mr. Norman
H. Horton and Mr. Jack W. Williamson on 13 October 1972. The
collection site was the Luten Mine of the Engelhard Minerals and
Chemicals Company (see Figure 3). It is located about 1 kilometer
east of Florida Highway 10 and 2 kilometers south of the Florida-
Georgia border in SE 1/4, NW 1/4, sec. 10, T.3N., R.3W., Gadsden
County, Florida. The Dogtown, Florida, 7.5' quadrangle map covers
this area.

The pit was being actively mined. After stripping the
overburden, the top 0.6 meter of the exposed clay was carefully
removed (Figure 4). The sample was processed in the specified
manner. This attapulgite was kindly donated by Engelhard Minerals
and Chemicals Company.

BIBLIOGRAPHY

Patterson, Sam H. (1974) Fuller's Earth and Other Mineral Resources
    of the Meigs-Attapulgus-Quincy District, Georgia and Florida,
    U.S.G.S. Professional Paper 828, p. 45.

Patterson, Sam H. and B.F. Buie (1974) Field Conference on Kaolin
    and Fuller's Earth, November 14-16, 1974, Guidebook 14,
    Georgia Geological Survey, p. 53.

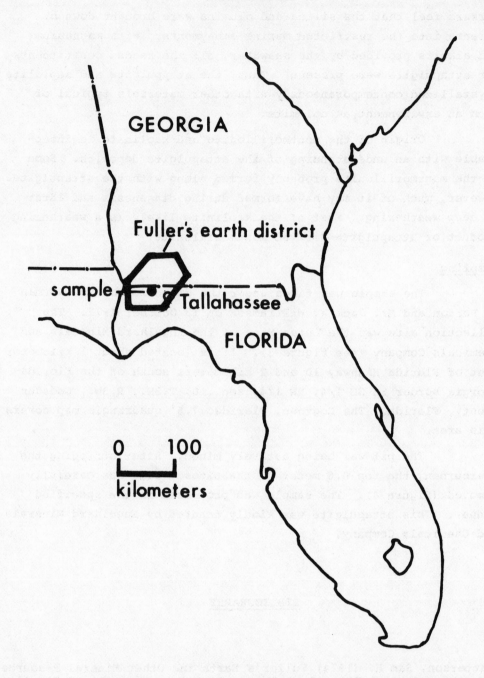

Figure 1.   Map of Florida and Georgia showing location of Fuller's
Earth district and attapulgite sample.

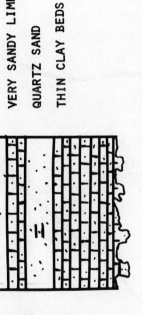

MICCOSUKEE FORMATION

BROWNISH CLAY
MEDIUM TO COARSE SAND
THIN KAOLINITE BEDS
CHANNEL-FILL
LENTICULAR QUARTZ GRAVEL UNITS

HAWTHORN FORMATION

FINE TO MEDIUM QUARTZ SAND AND CLAYEY SILT
FULLERS EARTH   SCATTERED VERY SANDY
LIMESTONE BEDS
SOME PHOSPHATE AND OPAL

TAMPA LIMESTONE

VERY SANDY LIMESTONE BEDS
QUARTZ SAND
THIN CLAY BEDS

Figure 2.   Geologic column of Fuller's Earth district (after
Patterson and Buie, 1974).

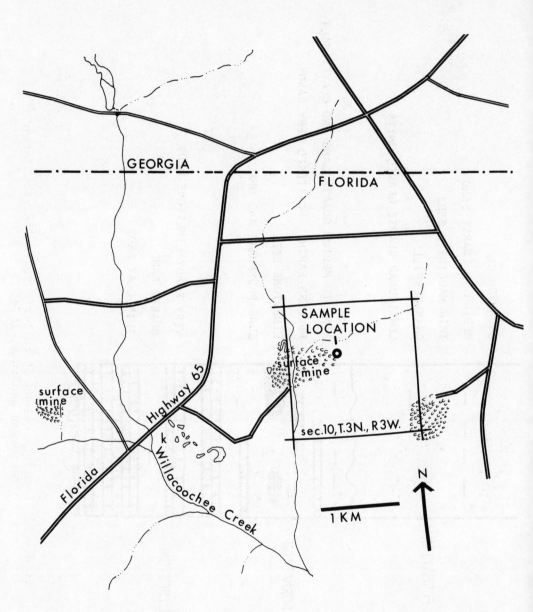

Figure 3.   Location of collection site of attapulgite sample.

Figure 4. Photograph of collection site of attapulgite sample.

CHEMICAL ANALYSIS

V. Gabis

INTRODUCTION

The results of chemical analysis provide information on the
mineralogical composition of the samples, both on the amount of
the primary mineral and on amount and nature of any other minerals
present in the sample. However, such interpretation of the chemical
analysis requires additional information, primarily crystallographic
information from x-ray analysis. Further clues are provided by the
results of dissolution experiments, thermal analysis, microscopic
and electron microscopic observations and spectral analysis.

Chemical analyses of the suite of CMS samples were performed
by two laboratories only. Chemical analysis of the OECD samples,
on the other hand, was specifically directed at an evaluation of
laboratory performance and of methods of analysis through a statistical analysis of a large number of analyses for each sample. This
goal was actually realized for seven of the samples only, for
which analyses were obtained from a minimum of eight, and a maximum
of nineteen laboratories. For the other minerals an insufficient
number of data were obtained and these are reported without further
analysis. A small number of analyses were performed on particle
size fractions below 2 μm, on calcium saturated samples and on
calcined samples, and also of trace elements on the total sample.
Only the latter will be reported.

CMS RESULTS

Two laboratories submitted complete results:

C.V.Clemency, Dept. of Geological Sciences, State University
  of New York at Buffalo, Buffalo,N.Y.

CMS CHEMICAL ANALYSIS

	KGa-1		KGa-2		SWy-1		STx-1		SAz-1	HCa-1	Syn-1		PFl-1	
	CL	MH	CL	MH	CL	MH	CL	MH	MH	MH	CL	MH	CL	MH
SiO_2	45.0	44.2	44.2	43.9	62.9	62.9	69.6	70.1	60.4	34.7	50.2	49.7	60.9	60.9
Al_2O_3	38.0	39.7	37.2	38.5	19.3	19.6	16.3	16.0	17.6	0.69	37.8	38.2	10.3	10.4
TiO_2	1.58	1.39	2.17	2.08	0.16	0.090	0.29	0.22	0.24	0.038	0.05	0.023	0.52	0.49
Fe_2O_3	0.26	0.13	1.14	0.98	3.85	3.35	1.17	0.65	1.42	0.02	0.03	0.02	3.33	2.98
FeO	0.02	0.08	0.05	0.15	0.12	0.32	0.04	0.15	0.08	0.25	0.00	--	0.16	0.40
MnO	0.00	0.002	0.00	--	0.01	0.006	0.01	0.009	0.099	0.008	0.00	--	0.03	0.058
MgO	0.02	0.03	0.04	0.03	2.80	3.05	3.56	3.69	6.46	15.3	0.19	0.014	10.40	10.2
CaO	0.02	--	0.04	--	1.80	1.68	1.73	1.59	2.82	23.4	0.02	--	1.86	1.98
Na_2O	0.01	0.013	0.02	<0.005	1.54	1.53	0.33	0.27	0.063	1.26	0.25	0.26	0.04	0.058
K_2O	0.04	0.050	0.02	0.065	0.56	0.53	0.15	0.078	0.19	0.13	0.02	<0.01	0.77	0.80
Li_2O										2.18		0.25		
P_2O_5	0.05	0.034	0.06	0.045	0.06	0.049	0.03	0.026	0.020	0.014	0.00	0.001	0.73	0.80
S	--	--	--	0.02	--	0.05	--	0.04	--	0.01	0.00	0.10	--	0.11
F (§)	(0.013)		(0.020)		(0.111)		(0.084)		(0.287)	(2.75)	(1.66)		(0.542)	
(°)										(2.60)		(0.76)		
ign.loss	14.31		14.24								11.28		10.33	
-550		12.6		12.6	5.10	1.59	7.50	3.32	7.54	1.20		8.75		8.66
550-1000		1.18		1.17		4.47		3.22	2.37	20.6		2.40		1.65
CO_2					1.33		0.16				0.12		0.32	
total	99.38	99.40	99.18	99.54	99.54	99.22	100.87	99.37	99.30	99.80	99.96	99.72	99.69	99.45

CL: Clemency   MH: Haydn   F(§): fluor by Thomas, not included in total; F(°): fluor included in ignition loss.

Haydn H. Murray, Dept. of Geology, Indiana University,
    Bloomington, Indiana.
A third laboratory submitted a set of analyses for fluor:
J. Thomas, Jr., Illinois State Geological Survey, Urbana, Ill.

All results are combined in the table.

OECD RESULTS

Reported analyses are presented in Tables (a).

For the five clays in the group of seven samples for which a
sufficient number of analyses were reported to allow a statistical
analysis of the results, the data were first recalculated on the basis
of a sample weight after calcination at $1000^{\circ}C$, using either the
data on loss on ignition, or those on contents of H_2O, CO_2, Cl and F.
The recalculated data are presented in Tables (b). The results for
talc and magnesite were not recalculated.Using only those data on
the seven samples which did not deviate more than one standard
deviation from the overall average, a most probable composition was
obtained, as recorded in the relevant Tables (c). These most probable
compositions are also summarized in a single Table. (page 150)

Results of trace element analyses are summarized in a separate
Table. (page 151)

OECD CHEMICAL ANALYSIS

01 (a)   Montmorillonite

	B1	B2	F2	F3	F5	F6	F9	F13	F19	D5
SiO_2	59.02	52.10	50.00	51.43	50.64	51.90	58.97	59.03	53.49	50.75
Al_2O_3	20.48	18.20	17.85	18.10	19.15	18.90	19.36	19.28	18.09	18.65
TiO_2	0.42	0.38	0.34	0.41		0.60	0.49	0.89	0.49	0.34
Fe_2O_3°	3.00	2.86	2.85	2.87	3.60	3.30	3.30	2.93	4.50	2.76
FeO °			0.18	0.06		0.30	0.22	0.19	0.15	0.27
MnO			<0.01	<0.01			0.01		0.01	0.02
MgO	4.60	4.05	4.05	3.64	3.68	5.30	4.33	4.58	4.00	3.74
CaO	2.03	1.75	2.48	2.58	2.10	3.00	2.56	2.38	2.78	2.41
Na_2O	1.62	1.74	1.55	1.44	1.48	1.50	1.71	1.19	1.42	1.38
K_2O	0.29	0.42	0.31	0.36	0.32	0.40	0.40	0.31	0.39	0.30
P_2O_5		0.06	<0.01	0.06			0.07		0.02	0.07
CO_2				0.93					0.60	0.46
SO_3		0.32								
F										
Cl										
H_2O 100C				12.90	9.32	7.90			6.81	12.54
1000C	8.09	17.90	20.00	4.49	9.33	7.30			6.48	6.44
ign.loss							8.16	8.99		
total	99.55	99.78	99.44	99.21	99.62	100.10	99.36	99.58	99.08	99.86
corr.: O=F,Cl										
corr.: total										

(°) Fe_2O_3 total Fe calculated as Fe_2O_3; FeO included in total

01 (a) continued

	D7	D15	D23	D28	P3	CH4	GB2	GB5	GB28
SiO_2	57.50	56.89	54.74	53.60	53.24	51.98	54.12	54.00	58.54
Al_2O_3	19.00	20.58	18.89	18.50	18.25	16.56	18.03	18.60	20.47
TiO_2	in Al	0.45	0.46	0.99	0.31	0.75	0.40	0.39	0.45
$Fe_2O_3°$	4.00	2.90	2.83	2.55	3.38	2.54	2.90	2.79	3.09
FeO °			0.21		0.24			0.15	
MnO			0.01		0.00			0.01	
MgO	4.60	4.12	4.04	1.99	4.28	4.41	4.27	3.90	4.27
CaO	1.60	2.24	2.48	2.44	2.37	2.38	2.44	2.34	2.79
Na_2O	2.50	1.40	1.26	1.35	1.11	0.93	1.36	1.56	1.55
K_2O	0.70	0.32	0.62	0.29	0.29	0.17	0.35	0.36	0.35
P_2O_5		0.07	0.06		0.00	0.18		0.07	0.10
CO_2			0.89	0.53	0.42				0.44
SO_3			0.40		0.26				
F								0.19	
Cl					0.25				
H_2O 100C			8.21					8.80	
1000C			4.21	17.50	16.67			6.97	7.74
ign.loss	9.90	10.85				18.77	14.81		
total	99.80	99.82	99.31	99.74	100.83	98.67	98.68	100.08	99.79
corr.: O=F,Cl					0.12			0.09	
corr.: total					100.71			99.99	

01 (b)   Montmorillonite

	B1	B2	F2	F3	F5	F6	F9	F13	F19	D5	D7	D15
SiO_2	64.53	63.63	62.95	63.57	62.54	61.13	64.65	65.17	62.79	63.11	63.96	63.94
Al_2O_3	22.39	22.23	22.47	22.38	23.65	22.26	21.23	21.28	21.23	23.19	21.13	23.13
TiO_2	0.46	0.46	0.43	0.51		0.71	0.54	0.98	0.58	0.42		0.51
Fe_2O_3	3.28	3.49	3.59	3.55	4.45	3.89	3.62	3.23	5.28	3.43	4.45	3.26
MnO							0.01		0.01	0.02		
MgO	5.03	4.95	5.10	4.50	4.54	6.24	4.75	5.06	4.70	4.65	5.12	4.63
CaO	2.22	2.14	3.12	3.19	2.59	3.53	2.81	2.63	3.26	3.00	1.78	2.52
Na_2O	1.77	2.13	1.95	1.78	1.83	1.77	1.87	1.31	1.67	1.72	2.78	1.57
K_2O	0.32	0.51	0.39	0.45	0.40	0.47	0.44	0.34	0.46	0.37	0.78	0.36
P_2O_5		0.07		0.07			0.08		0.02	0.09		0.08
SO_3		0.39										
total	100.00	100.00	100.00	100.00	100.00	100.00	100.00	100.00	100.00	100.00	100.00	100.00

O1 (b) continued   O1 (c)

	D23	D28	P3	CH4	GB2	GB5	GB28	n	x̄	σ	x̄ corr.
SiO_2	63.81	65.60	63.76	65.05	64.52	64.27	63.97	19	63.86	1.05	63.84
Al_2O_3	22.02	22.64	21.86	20.73	21.50	22.14	22.37	19	22.10	0.48	22.24
TiO_2	0.54	1.21	0.37	0.94	0.48	0.46	0.49	17	0.59	0.23	0.50
Fe_2O_3	3.30	3.12	4.05	3.18	3.46	3.32	3.38	19	3.65	0.55	3.45
MnO	0.01					0.01		5	0.01		
MgO	4.71	2.44	5.15	5.52	5.09	4.64	4.67	19	4.82	0.71	4.87
CaO	2.89	2.99	2.84	2.98	2.91	2.79	3.05	19	2.80	0.42	2.88
Na_2O	1.47	1.65	1.33	1.16	1.62	1.86	1.69	19	1.73	0.34	1.73
K_2O	0.72	0.35	0.35	0.21	0.42	0.43	0.38	19	0.43	0.13	0.40
P_2O_5	0.07			0.23		0.08		9	0.09	0.05	0.08
SO_3	0.46		0.31					2	0.35		
total	100.00	100.00	100.00	100.00	100.00	100.00	100.00				

02 (a)   Laponite

	B1	B2	F3	F5	F9	F19	D5	GB28
SiO_2	60.56	55.85	56.36	54.60	60.56	54.07	53.96	61.18
Al_2O_3	0.36		0.55	0.00	0.10	0.15	0.47	0.46
TiO_2	0.03	0.03	0.02		0.03	0.00	0.02	0.03
$Fe_2O_3°$	0.05	0.03	0.08	0.00	0.10	0.00	0.06	0.05
FeO °					0.04		<0.01	
MnO			0.01		0.01	0.00	<0.01	
MgO	27.25	24.00	24.46	24.30	26.59	26.00	24.03	26.32
CaO	0.26	0.31	0.21	0.60	0.50	0.30	0.27	0.22
Na_2O	2.90	3.15	2.72	2.88	2.79	2.61	2.57	2.90
K_2O	0.03	0.05	0.02	0.04	0.05	0.03	0.03	0.03
P_2O_5			0.01		0.02	0.02		
CO_2		0.56	0.74				0.33	0.30
SO_3		0.09						
F								
Li_2O		0.94		0.90	0.50		0.92	0.96
H_2O 100C						6.89	10.66	
1000C	8.46	14.66	14.87			7.93	7.16	7.32
ign.loss				16.00	8.56			
total	99.90	99.67	100.05	99.12	99.81	98.26	100.48	99.77
corr.: $O=F,Cl$								
corr.: total								

(°) Fe_2O_3 total Fe calculated as Fe_2O_3; FeO included in total

O2 (b) Laponite

	B1	B2	F3	F5	F9	F19	D5	GB28	O2 (c) n	$\bar{x}$		$\bar{x}$ corr
SiO_2	66.24	66.12	66.76	65.53	66.37	65.00	65.54	66.40	8	66.00	0.57	66.03
Al_2O_3	0.39	0.00	0.65	0.00	0.11	0.18	0.57	0.50	8	0.30	0.26	0.30
TiO_2	0.03	0.04	0.02		0.03	0.00	0.02	0.03	7	0.02	0.01	0.02
Fe_2O_3	0.06	0.04	0.09	0.00	0.11	0.00	0.07	0.05	8	0.05	0.04	0.06
MnO			0.01		0.01	0.00	0.00		4	0.01		
MgO	29.80	28.42	28.97	29.16	29.14	31.26	29.19	28.56	8	29.31	0.89	29.03
CaO	0.28	0.37	0.25	0.72	0.55	0.36	0.33	0.24	8	0.39	0.17	0.34
Na_2O	3.17	3.73	3.22	3.46	3.06	3.14	3.12	3.15	8	3.26	0.23	3.19
K_2O	0.03	0.06	0.02	0.05	0.05	0.04	0.04	0.03	8	0.04	0.01	0.04
P_2O_5			0.01		0.02	0.02			3	0.02		
SO_3		0.11							1			
Li_2O		1.11		1.08	0.55		1.12	1.04	5	0.98		
total	100.00	100.00	100.00	100.00	100.00	100.00	100.00	100.00				

03 (a)   Kaolinite (China Clay)

	B1	B2	F3	F5	F6	F9	F13	F19	D5
SiO_2	46.72	46.90	46.47	45.82	43.75	46.56	47.58	44.52	45.94
Al_2O_3	38.46	38.00	38.35	38.80	39.10	37.36	36.93	38.25	39.24
TiO_2	0.03	0.03	0.02	in Al	0.00	0.03	0.08	0.00	0.03
Fe_2O_3 °	0.43	0.45	0.44	0.60	0.61	1.02	0.72	1.56	0.46
FeO °			0.07		0.01	0.06			
MnO						0.01		0.00	<0.01
MgO	0.04	0.16	0.10	0.17	0.00	0.40	0.19	0.00	0.13
CaO	0.00	0.08	0.06	0.00	1.70	0.11	0.19	0.59	0.07
Na_2O	0.07	0.07	0.04	0.04	0.10	0.08	0.02	0.06	0.05
K_2O	1.28	1.18	1.22	1.14	1.30	1.26	1.00	1.33	1.19
P_2O_5		0.09	0.10			0.10	0.28	0.08	0.10
CO_2			0.28					0.00	<0.01
SO_3									
F		0.11							
Cl									
H_2O 100C		0.27	0.26		0.10			0.08	0.26
1000C	12.89	13.00	12.20	12.60	13.00			13.06	12.77
ign.loss						12.69	12.93		
total	99.92	100.34	99.54	99.17	99.66	99.62	99.92	99.53	100.24
corr.: O=F,Cl		0.05							
corr.: total		100.29							

(°) Fe_2O_3 total Fe calculated as Fe_2O_3; FeO included in total

03 (a) continued

	D7	D15	D28	P3	CH4	GB2	GB5	GB19	GB28
SiO_2	44.10	46.62	46.90	46.42	46.74	46.30	46.75	46.69	46.37
Al_2O_3	40.40	37.60	38.50	38.10	38.37	37.86	38.00	37.63	38.21
TiO_2	in Al	0.03	0.06	0.00	0.03	0.05	0.04	0.07	0.06
Fe_2O_3 °	1.10	0.46	0.41	0.56	0.49	0.54	0.46	0.43	0.44
FeO °				0.19			0.06		
MnO				0.00			<0.01	0.01	
MgO	0.20	0.20	0.08	0.18	0.14	0.18	0.17	0.09	0.16
CaO	0.40	0.08	0.22	0.20	0.04	0.12	0.08	0.60	0.09
Na_2O	0.50	0.10	0.07	0.11	0.06	0.06	0.09	0.07	0.05
K_2O	2.50	1.24	1.19	1.11	1.21	1.12	1.20	1.13	1.20
P_2O_5		0.33		0.00	0.05	0.32	0.10	0.08	0.10
CO_2				0.10	0.10		<0.01	0.04	0.00
SO_3				0.04			<0.02	0.08	
F						0.13	0.10	0.08	
Cl							<0.01	0.01	
H_2O 100C				0.14			0.17		
1000C				13.10	12.76	13.06	12.92	12.93	
ign.loss	11.10	12.88	12.40						13.00
total	100.30	99.54	99.83	100.25	99.99	99.74	100.08	99.94	99.68
corr.:						0.06	0.05	0.04	
O=F,Cl									
corr.						99.68	100.03	99.90	
total									

O3 (b)   Kaolinite (China Clay)

	B1	B2	F3	F5	F6	F9	F13	F19	D5	D7	D15
SiO_2	53.69	53.94	53.54	52.93	50.55	53.56	54.70	51.53	52.69	49.44	53.80
Al_2O_3	44.19	43.70	44.18	44.82	45.17	42.98	42.45	44.28	44.99	45.29	43.39
TiO_2	0.03	0.03	0.02		0.00	0.03	0.09	0.00	0.03		0.03
Fe_2O_3	0.49	0.52	0.51	0.69	0.70	1.17	0.83	1.81	0.53	1.23	0.53
MnO						0.01					
MgO	0.05	0.18	0.12	0.19	0.00	0.46	0.22	0.00	0.15	0.23	0.23
CaO	0.00	0.09	0.07	0.00	1.96	0.13	0.22	0.68	0.08	0.45	0.09
Na_2O	0.08	0.08	0.05	0.05	0.12	0.09	0.02	0.07	0.06	0.56	0.12
K_2O	1.47	1.36	1.40	1.32	1.50	1.45	1.15	1.54	1.36	2.80	1.43
P_2O_5	0.11	0.10	0.11			0.12	0.32	0.09	0.11		0.38
SO_3											
total	100.00	100.00	100.00	100.00	100.00	100.00	100.00	100.00	100.00	100.00	100.00

O3 (b) continued | | | | | | | O3 (c)

	D28	P3	CH4	GB2	GB5	GB19	GB28	n	x̄	σ	x̄ corr.
SiO_2	53.64	53.46	53.65	53.50	53.80	53.75	53.50	18	53.10	1.31	53.53
Al_2O_3	44.04	43.88	44.03	43.74	43.73	43.31	44.08	18	44.06	0.78	43.95
TiO_2	0.07	0.00	0.03	0.06	0.05	0.08	0.07	16	0.04	0.02	0.03
Fe_2O_3	0.47	0.65	0.56	0.62	0.53	0.50	0.51	18	0.71	0.35	0.58
MnO						0.01		2	0.01		
MgO	0.09	0.21	0.16	0.21	0.20	0.10	0.18	18	0.16	0.11	0.17
CaO	0.25	0.23	0.05	0.14	0.09	0.69	0.10	18	0.30	0.46	0.20
Na_2O	0.08	0.13	0.07	0.07	0.10	0.08	0.06	18	0.11	0.11	0.08
K_2O	1.36	1.28	1.39	1.29	1.38	1.30	1.38	18	1.45	0.35	1.37
P_2O_5		0.11	0.06	0.37	0.12	0.09	0.12	13	0.16	0.11	0.10
SO_3		0.05				0.09					
total	100.00	100.00	100.00	100.00	100.00	100.00	100.00				

04 (a)   Attapulgite

	B1	F2	F5	F6	F9	F13	F19	D7	D28	GB5	GB28
SiO_2	68.84	63.94	63.50	63.90	68.15	66.32	62.68	59.50	65.80	66.05	68.81
Al_2O_3	8.63	7.70	8.00	9.80	7.75	9.92	8.00	11.50	8.93	8.22	8.57
TiO_2	0.40	0.55	0.80	0.80	0.52	0.94	0.70	in Al	0.79	0.57	0.57
Fe_2O_3 °	2.76	2.53	3.40	2.60	3.18	2.81	3.52	3.50	2.50	2.62	2.74
FeO °				0.40	0.06		0.15				
MnO		0.10			0.09		0.08			0.08	
MgO	7.62	7.35	7.50	4.90	7.88	7.48	7.06	11.00	6.38	7.19	7.44
CaO	1.40	1.85	1.46	3.00	1.88	1.70	2.10	2.50	1.76	1.64	1.80
Na_2O	0.14	0.17	0.13	0.10	0.15	0.09	0.11	0.20	0.12	0.13	0.13
K_2O	0.62	0.65	0.59	0.70	0.70	0.55	0.75	0.90	0.61	0.65	0.64
P_2O_5					0.02		0.00				
CO_2							0.58		1.41		1.77
SO_3											
F		0.09								0.09	
Cl											
H_2O 100C			5.85	4.60			4.39			3.83	
1000C			10.10	9.80			8.91		11.50	9.03	7.66
ign.loss	9.56	15.90			9.64	9.95		10.30			
total	99.97	100.83	100.53	100.20	99.96	99.76	98.88	99.40	99.80	100.10	100.13
corr.: O=F,Cl		0.04								0.04	
corr.: total		100.79								100.06	

(°) Fe_2O_3 total Fe calculated as Fe_2O_3; FeO included in total

O4 (b) Attapulgite

O4 (c)

	B1	F2	F5	F6	F9	F13	F19	D7	D28	GB5	GB28	n	x̄	σ	x̄ corr.
SiO_2	76.14	75.36	75.08	74.48	75.45	73.84	73.75	66.78	75.72	75.79	75.87	11	74.39	2.65	75.15
Al_2O_3	9.55	9.08	9.47	11.41	8.58	11.05	9.41	12.90	10.28	9.43	9.45	11	10.05	1.26	9.72
TiO_2	0.44	0.65		0.93	0.58	1.05	0.82		0.91	0.65	0.63	9	0.74	0.20	0.74
Fe_2O_3	3.05	2.98	4.02	3.03	3.52	3.13	4.14	3.93	2.88	3.01	3.02	11	3.34	0.47	3.08
MnO		0.12			0.10		0.09			0.09		4	0.10		
MgO	8.43	8.66	8.87	5.71	8.72	8.33	8.31	12.35	7.34	8.25	8.20	11	8.47	1.56	8.35
CaO	1.55	2.18	1.73	3.50	2.08	1.89	2.47	2.81	2.03	1.88	1.98	11	2.19	0.56	2.03
Na_2O	0.15	0.20	0.15	0.12	0.17	0.10	0.13	0.22	0.14	0.15	0.14	11	0.15	0.03	0.14
K_2O	0.69	0.77	0.70	0.82	0.78	0.61	0.88	1.01	0.70	0.75	0.71	11	0.76	0.11	0.74
P_2O_5					0.02							1	0.02		
SO_3															
total	100.00	100.00	100.00	100.00	100.00	100.00	100.00	100.00	100.00	100.00	100.00				

05 (a) Illite

	B9	F2	F5	F6	F9	F13	F19	D7	D15	D23	D28	CH4	GB28
SiO_2	52.72	50.40	45.68	49.50	42.84	52.66	47.42	50.79	51.80	52.28	47.20	50.04	51.70
Al_2O_3	20.73	19.40	19.00	20.00	16.29	20.30	18.00	22.97	21.50	20.31	18.00	23.60	20.40
TiO_2	0.77	0.95	in Al	0.60	0.67	1.08	0.87	in Al	0.75	0.75	1.25	0.71	0.77
Fe_2O_3 °	6.69	6.25	6.40	7.00	6.50	6.79	7.10	7.73	6.70	6.66	2.92	5.29	6.49
FeO °				0.50	0.36		0.45			0.59			
MnO		0.03			0.01		0.04			0.04			
MgO	3.55	3.80	3.24	3.30	3.10	3.82	2.82	3.13	3.68	3.68	2.73	3.37	3.72
CaO	1.38	1.15	1.01	2.00	1.03	1.10	1.96	2.28	0.83	1.07	1.19	0.97	1.19
Na_2O	0.36	0.37	0.31	0.40	0.38	0.29	0.29	0.38	0.23	0.28	0.28	0.72	0.49
K_2O	8.94	7.85	7.10	8.20	6.87	7.91	8.15	8.01	7.25	8.49	6.85	8.41	8.22
P_2O_5					0.40		0.04		0.40	0.43			
CO_2							0.50			0.18			0.14
SO_3										0.05			
H_2O 100C				3.10	16.76		8.36			2.19		0.95	
H_2O 1000C				5.60	4.29		4.51	4.30	6.88	3.49		5.84	
ign.loss	4.96	9.00	17.80			5.52					19.00		6.78
total	100.10	99.20	100.54	99.70	99.14	99.47	100.06	99.49	100.02	99.90	99.42	99.90	99.76

(°) Fe_2O_3 total Fe calculated as Fe_2O_3; FeO included in total

05 (b) Illite 05 (c)

	B1	F2	F5	F6	F9	F13	F19	D7	D15	D23	D28	CH4	GR28	n	x̄	σ	x̄ corr.
SiO_2	55.41	55.88	55.21	54.39	54.86	56.05	54.70	53.25	55.62	55.62	58.70	53.75	55.60	13	55.31	1.32	54.58
Al_2O_3	21.79	21.51	22.96	21.98	20.86	21.61	20.76	24.13	21.61	21.61	22.38	25.35	21.94	13	22.19	1.29	21.93
TiO_2	0.81	1.05		0.66	0.86	1.15	1.10		0.80	0.80	1.55	0.76	0.83	11	0.93	0.25	0.90
Fe_2O_3	7.03	6.93	7.74	7.69	8.32	7.23	8.19	8.12	7.08	7.09	3.63	5.68	6.98	13	7.05	1.24	7.41
MnO		0.03			0.01		0.05		0.04	0.04				5	0.03		
MgO	3.73	4.21	3.92	3.63	3.97	4.06	3.25	3.29	3.92	3.91	3.39	3.62	4.00	13	3.76	0.97	3.76
CaO	1.45	1.28	1.22	2.20	1.32	1.17	2.26	2.40	1.14	1.14	1.48	1.04	1.28	13	1.49	0.47	1.25
Na_2O	0.38	0.41	0.37	0.44	0.49	0.31	0.34	0.40	0.30	0.30	0.35	0.77	0.53	13	0.41	0.13	0.39
K_2O	9.40	8.70	8.58	9.01	8.80	8.42	9.40	8.41	9.03	9.03	8.52	9.03	8.84	13	8.86	0.33	8.88
P_2O_5					0.51		0.05		0.46	0.46				4	0.48		
SO_3																	
total	100.00	100.00	100.00	100.00	100.00	100.00	100.00	100.00	100.00	100.00	100.00	100.00	100.00				

06 (a) Chrysotile

	B1	F9	D16	D29	GB36
SiO_2	41.19	41.65			41.33
Al_2O_3	0.82	0.83		0.76	0.52
TiO_2	0.02	0.05		0.01	in Al
Fe_2O_3 °	1.54	2.04	1.79	1.91	1.94
FeO °		0.14	0.20	0.13	0.25
MnO		0.04		0.04	0.04
MgO	43.20	41.60	41.80	41.51	42.36
CaO	0.16	0.26	0.06	0.04	0.25
Na_2O	0.07	0.07	0.01	0.02	0.02
K_2O	0.02	0.03	0.01		0.01
P_2O_5		0.08			in Al
CO_2					
SO_3					
Cl					
H_2O 110C					
1000C	13.84		13.82		
ign.loss		13.45		13.58	13.94
total	100.86	99.99			100.41

(°) Fe_2O_3 total Fe calculated as Fe_2O_3;
    FeO included in total

07 (a)   Crocidolite

	B1	F5	F9	I1	D16	D29
SiO_2	48.71	48.58	48.00	48.16		
Al_2O_3	0.20	0.44	1.02	0.27		0.20
TiO_2	0.02		0.05	0.03		
Fe_2O_3 °	40.00	42.35	39.00	41.16	40.00	43.19
FeO °	21.59	18.65	19.61	18.71	20.00	20.24
MnO			0.14	0.15		0.13
MgO	2.10	0.67	2.21	2.32	2.02	2.18
CaO	1.08	2.40	1.17	1.11	1.15	0.99
Na_2O	5.93	5.20	6.00	5.16	5.40	5.61
K_2O	0.10	0.09	0.09	0.07	0.08	
P_2O_5			0.04		0.01	
CO_2						
SO_3						
H_2O 110C			0.76			
1000C	1.19		3.38			
ign.loss				1.12	2.42	1.55
total	99.33	99.73	101.86	100.55		

(°) Fe_2O_3 total calculated as Fe_2O_3; FeO included
         in total Fe

08 (a) Talc

	B1	B2	F3	F5	F6	F9	F13	D16	D23	D29	I1	08 (c) n	$\bar{x}$	σ	$\bar{x}$ corr.
SiO_2	61.08	60.85	60.83	60.00	58.60	59.72	59.91		62.32		60.13	9	60.27	0.84	60.36
Al_2O_3	0.66	0.67	0.83	0.40	0.80	0.46	0.86		0.94	1.53	2.29	10	0.94	0.56	0.70
TiO_2	0.02	0.03	0.03		0.30	0.02	0.50		0.06	0.05		8	0.12	0.18	0.07
Fe_2O_3°	0.82	0.85	0.91	0.92	1.00	1.04	1.04	0.83	0.87	0.98	1.18	11	0.95	0.11	0.95
FeO°			0.63		0.80	0.53			0.71	0.76	0.35				
MnO			0.01			0.01			0.01		0.08	4	0.03		
MgO	32.22	31.40	31.35	31.40	31.70	31.86	31.07	31.80	31.02	31.12	30.91	11	31.44	0.41	31.41
CaO	0.16	0.46	0.42		1.80	0.69	0.43	0.51	0.42	0.42	0.74	10	0.61	0.45	0.47
Na_2O	0.55		0.02	0.05		0.07	0.04	0.04	0.01	0.03	0.03	8	0.10	0.18	0.04
K_2O	0.21		0.02	0.03		0.05	0.01	0.01	0.06	0.02	0.02	7	0.06	0.07	0.03
P_2O_5			0.17			0.18		0.15	0.16			4	0.15		
CO_2			1.02						0.86						
SO_3									0.13						
H_2O 100C					0.40				0.05		0.06				
1000C				5.60	5.40				4.67		4.87				
ign.loss	5.62	5.60	4.62			5.74	5.68	4.54		5.47					
total	101.34		100.16	98.70	99.90										

(°) Fe_2O_3 total Fe calculated as Fe_2O_3; FeO included in total

	10 (a) Gibbsite			13 (a) Gypsum					
	B1	B2	F5	B1	B3	F2	F5	EIR1	GB28
SiO_2	0.64	0.05	0.00	1.73	1.27	1.25	1.10	0.94	1.67
Al_2O_3	65.30	65.45	65.20	0.27	0.14				0.26
TiO_2	0.00			0.01			} 0.44	} 0.45	0.03
Fe_2O_3 °	0.04	0.02		0.10	0.10		}	}	0.08
FeO						0.07			
MnO									
MgO	0.01			0.50	0.66	0.60	1.80	0.23	0.52
CaO	0.00			40.46	32.99	33.05	32.00	33.60	40.09
Na_2O	0.36	0.38	0.37	0.14	0.12		0.09	0.40	0.05
K_2O	0.03		0.00	0.08	0.00		0.08	0.06	0.06
P_2O_5									
CO_2						1.25		2.25	3.06
SO_3				52.53	42.40	42.65	42.30	43.40	52.59
Cl								0.17	
H_2O 110C								18.77	
1000C	33.13	33.60							
ign.loss			34.30	4.06	22.17	20.65	21.76		1.54
total	98.91	99.50	99.87	99.88	100.13	99.52	99.57	100.27	99.95

(°) Fe_2O_3 total calculated as Fe_2O_3; FeO included in total Fe

11 (a) Magnesite

	A1	B1	B2	B3	F2	F5	F13	I1	EIR1	11 (c) n	$\bar{x}$	σ	$\bar{x}$ corr.
SiO_2	4.04	5.76	5.76	5.66	5.30	3.70	5.43	3.90	4.64	9	4.91	0.85	5.43
Al_2O_3	0.94	1.06	0.95	1.16	0.95	1.05	0.88	1.00	0.58	9	0.95	0.16	0.98
TiO_2	0.03	0.04	0.03	0.05	0.11		0.21		0.08	7	0.08	0.06	0.06
Fe_2O_3 °	1.56	1.30	1.49	1.36	1.45	1.84	1.41	1.51	1.78	9	1.52	0.18	1.44
FeO °	0.58		0.68		0.74								
MnO	0.06		0.02	0.04	0.05			0.06		5	0.05		
MgO	43.05	42.39	42.25	42.33	41.95	43.20	42.56	42.96	41.69	9	42.49	0.51	42.50
CaO	2.16	2.73	2.34	1.97	2.75	3.26	2.67	2.69	2.45	9	2.56	0.37	2.61
Na_2O		0.19	0.05	0.12		0.10	0.25	0.07	1.57	7	0.33	0.55	0.13
K_2O		0.11	0.11			0.10	0.10	0.09	0.05	6	0.09		
P_2O_5													
CO_2					0.80								
SO_3													
H_2O 110C													
1000C									47.26				
ign.loss	48.07	46.51	46.70	47.07	46.43	46.50	46.06						
total	99.91	100.09				99.75	99.57						

(°) Fe_2O_3 total calculated as Fe_2O_3; FeO included in total Fe

12 (a)   Calcite

	B1	B2	B3	F2	F3	F5	EIR1	GB5
SiO_2	0.15	0.08	0.18	<0.05	0.21	0.00		0.11
Al_2O_3	0.06		0.03		0.57	)	)	0.01
TiO_2	0.00				0.00	}0.14	}0.50	
Fe_2O_3 °	0.06	0.06	0.05	<0.05	0.09	)	)	0.06
FeO °					0.05			
MnO								
MgO	0.29	0.33	0.30	0.40	0.27	0.80	0.00	0.28
CaO	55.58	55.47	55.06	55.42	55.19	54.00	54.60	55.38
Na_2O	0.07		0.03		0.24	0.07	0.94	0.03
K_2O	0.03		0.00		0.08	0.02	0.06	0.01
P_2O_5							0.01	
CO_2				43.60				
SO_3			0.08					
H_2O 110C								
H_2O1000C								
ign.loss	43.39	43.65	43.79		43.48	45.40	44.06	43.86
total	99.63	99.59	99.52	99.52	100.13	100.43	100.17	99.74

(°) Fe_2O_3 total calculated as Fe_2O_3; FeO included in total Fe

Most probable composition of seven samples

	01	02	03	04	05	08	11
SiO_2	63.84	66.03	53.53	75.15	54.58	60.36	5.43
Al_2O_3	22.24	0.30	43.95	9.72	21.93	0.70	0.98
TiO_2	0.50	0.02	0.03	0.74	0.90	0.07	0.06
Fe_2O_3	3.45	0.06	0.58	3.08	7.41	0.95	1.44
MnO	0.01	0.01	0.01	0.10	0.03	0.03	0.05
MgO	4.87	29.03	0.17	8.35	3.76	31.41	42.50
CaO	2.88	0.34	0.20	2.03	1.25	0.47	2.61
Na_2O	1.73	3.19	0.08	0.14	0.39	0.04	0.13
K_2O	0.40	0.04	1.37	0.74	8.88	0.03	0.09
P_2O_5	0.08	0.02	0.10	0.02	0.48	0.15	
SO_3	0.35						
Li_2O		0.98					

Trace Elements   ppm

	F2 01	D28 01	D5 02	D5 03	F2 04	D28 04	F2 05	D28 05	D29 06	GB36 06	D29 07	D29 08	GB36 08	A1 11	F2 11	D29 11	D29 12	F2 13	D29 13
Ni	17	39	<10	10	22	100	40	118		1400	<10	14		7	7	<10	<10	<5	<10
Ag	<1				<1		<1								<1			<1	
Sn	11			50	13		20								120			<2	
Mn	74	223		10	932	770	260	287				41	30		322			<10	
Ge	<6				<6		<6		2.1						<6			<6	
Sr	185				30		80				15			96	207			1967	
Ba	1250				930		323							670	760			49	
Cr	8	21			39	36	72	54		545	65	21	8		12			25	
Yb	<2		<10		<2		<2								<2			<2	
Sc	<2				7		9			11			1.3		9			<2	
Y	<10			10	<10		<10								<10			<10	
Co	<5		<10	20	11		10						4		8			<5	
Cu	21	22	<10		11	11	67	40		50				3	13			<3	
V	17				56		55							17	<10			<10	
Mo	<7				<7		<7							3	<7			<7	
Bi	<3				<3		<3								<3			<3	
Ga	18			20	12		26								5			<3	
Pb	21	30			18	30	24	28						11	<6			<6	
B	40			151	42		159								<20			<20	
Be	<3				<3		3								<3			<3	
Cd	<6				<6		<6								<6			<6	
Zn	109	114	100	200	24	70	112	136							<20			<20	
Pt	<10																		
Au	<10																		
Nb	<10																		
La	<50																		
Sb	<500																		
As	<500																		
Tl	<500																		

OECD COMMENTS

V.Gabis, Département des Sciences de la Terre,
                Université d'Orléans, France

A cursory examination of the results shows that there is
a considerable spread in the data. This spread may be due to

(a) differences in the state of hydration of the samples as
    a result of differences in storage conditions in the various
    laboratories;

(b) differences in presentation of the analytical results: they
    are either based on the weight of the wet sample or on that
    of the sample dried at $110^{\circ}C$;

(c) the application of a variety of analytical methods. Standard-
    ization of methods and procedures would undoubtedly reduce the
spread of the data.

A comparative study was made of the relative reliability of
the analytical methods which were used by the participants. These
methods may be grouped in the following categories:

A Atomic absorption

C Colorimetric methods

G Gravimetric methods

F X-ray fluorescence

P. Flame photometry

V Volumetric methods

The following table shows the percentages of analyses falling
within one standard deviation from the average, together with the
total number of analyses performed (in brackets).

SiO_2	G: 68% (59)	C: 86% (7)			C>G
Al_2O_3	G: 68% (37)	C: 75% (12)	V: 65% (17)		C>G=V
TiO_2	C: 77% (48)	F: 100% (3)			
Fe_2O_3	G: 50% (1)	C: 92% (37)	V: 50% (28)	F: 100% (1)	C>V
MgO	G: 83% (18)	C: 100% (4)	V: 81% (37)	A: 90% (10)	C>A>G=V
CaO	G: 54% (11)	C: 89% (9)	V: 83% (42)	A: 71% (7)	C>V>A>G
			P: 100% (1)		
Na_2O	A: 86% (7)	P: 85% (60)			A=P
K_2O	A: 67% (6)	P: 80% (57)			P>A
P_2O_5	G: 100% (1)	C: 82% (11)			

The last column indicates the relative reliability of the methods, omitting those which are little used.

As far as the human factor is concerned, it is clear that some analysts are systematically better than others: in one laboratory, out of 24 analyses performed, the analyst gives 24 results which are within one standard deviation; for another laboratory the analyst gives only 2 results within the standard deviation out of 6 analyses performed.

X RAY FLUORESCENCE SPECTROSCOPY

K. Jasmund

INTRODUCTION

Because of its relative simplicity and high precision, X-ray
fluorescence spectroscopy has been widely applied to silicate
analysis in the past decades. It is particularly useful for certain
elements which are difficult for laboratories to determine chemically,
but it is limited in that elements of low atomic number, such as Na,
Mg, and Al cannot be determined with a high degree of accuracy. Some
other elements, particularly those which are used as anode materials,
are also difficult to determine accurately. However, the method can
be applied to Fe, Ti, Ca, Pb, Zn, Zr, S, Cl, etc. with a good degree
of accuracy provided adequate precautions are taken. Most currently
available instruments permit the analysis of all elements down to
the atomic number of 10 in a concentration range from 100% to a
few ppm.

The principle of the method is briefly as follows: All elements
present in a sample are excited by a primary X-ray beam to emit their
own characteristic radiation. This emitted fluorescence radiation is
dispersed by diffraction on suitable crystals so that the characteristic
lines of elements can be recorded by proportional or scintillation
counters.

The application of X-ray fluorescence spectroscopy to the analysis
of geological samples (rocks, minerals, and mineral mixtures) has
shown that particular care has to be taken with two aspects, namely
sample preparation and the mutual interference between elements
(matrix effects). This is especially necessary if high accuracy
is required, therefore, special attention should be paid to the points
discussed below.

155

Sample preparation

Only a thin surface layer contributes to the radiation being analyzed. Consequently, when the composition of the surface layer is not representative of that of the bulk material, the fluorescent intensities of elements depend in a complex way on the particle shape and particle size distribution of the phases present. The possible error increases as the wavelength of the radiation selected for the analysis, and the difference in mass absorption coefficients of the phases increase. These effects are most easily eliminated by fusing the samples with $Na_2B_4O_7$. On quenching.the melt becomes a glass in which the elements are homogeneously distributed. For the fusion, crucibles of an alloy of Pt with 5% Au have proved to be very satisfactory because the sample contracts to a ball on cooling and can be easily removed from the crucible. The glass is then finely ground and pressed into a sample holder.

Matrix effects

The intensity of the X-ray radiation of an element depends in a complicated way on absorption or enhancement by other elements present in the sample.

One way to eliminate these effects is the use of internal standards. In this method, the intensity of a fluorescence line of the element to be determined is compared with that of a line from a reference element (originally absent) which has been added to the sample in a known amount. If the lines being compared behave similarly in as far as absorption and enhancement by other elements is concerned, a linear relationship exists between the intensity ratio of the lines and the concentration of the element being determined. The following line pairs, for example, have proved to be satisfactory:

Element to be determined	Internal standard
Ba K_α	Cd K_α
Fe K_α	Sm $L_{\beta 1}$
Ca K_α	Sc K_α

The internal standard is fused together with the sample in $Na_2S_4O_7$.

Another technique which is useful in certain circumstances is the method of incremental addition of small amounts of the element being determined (often known as the "spiking" technique). If the amounts added are small, the matrix effects should not be appreciably affected.

Precision

Intensity measurements in X-ray fluorescence spectroscopy are governed by the laws of radioactive decay. Apart from the effects of sample preparation and the particular conditions in the apparatus, the precision of the method is in the first instance given by the "counting error". The relative standard deviation is given by $\varepsilon = 1/\sqrt{N}$ or, expressed as a percentage: ε x100.

This shows that the precision of the intensity measurement can be improved by increasing the number of impulses counted. Normally, 100.000 impulses will be satisfactory, because the percentage relative standard deviation will then be below 0.8% for 99 out of 100 measurements.

Accuracy

X-ray fluorescence is not an absolute analytical method in itself. The concentration of an element is determined by comparing the intensity of a line of an element in the sample with the intensity of the same line in the reference sample containing the element in a known amount. In the internal standard method, calibration curves are prepared by plotting the quotient

$$\frac{\text{intensity for element}}{\text{intensity for internal standard}}$$

of reference samples against the known amount of the element in these samples. The accuracy of the method is therefore as good as the accuracy of concentration in the reference sample, provided the internal standard has been carefully selected. Reference samples of the U.S.National Bureau of Standards and British Chemical Standards have proved to be suitablefor the preparation of calibration curves.

Equipment of the spectrograph

For the analysis of elements with atomic numbers smaller than 22, a vacuum spectrograph must be used because the radiation of these elements is strongly absorbed by air. The spectrograph should be equipped with the following items:

2 X-ray tubes: W for heavy elements

      Cr for light elements

Crystals: LiF for elements >Ca

      PE for elements Ca-Al

      ADP or gypsum for elements <Al

2 detectors: scintillation counter for elements >Fe

proportional counter for elements <Fe

References

Birks,L.S. X-ray spectrochemical analysis. Interscience, New York,1959

Liebnafsky,H.A., Pfeiffer,H.G., Winslow,E.H. and Zemany,P.D.,
    X-ray absorption and emission in anlytical chemistry. Wiley,
    New York, 1960.

Norrish,K. and Chappell,B.W. "X-ray fluorescence spectrography",
in: Physical Methods in Determinative Mineralogy, J.Zussman, Ed.
Academic, London, 1967.

OECD RESULTS

In addition to a few incidental determinations as part of the
chemical analysis as reported in that section, X-ray fluorescence
analyses (including those of trace elements) were carried out by
nine laboratories on 10 of the minerals. The results are collected
in the tables (a) for the principal elements, and in table (b) for
the trace elements. For comparison, the results of chemical analyses
reported before, are included in tables (a).

OECD COMMENTS

The interlaboratory comparison of the data for montmorillonite, illite
and attapulgite is rather difficult. Since these materials lose some
water below 110°C, pretreatment prior to analysis is important. Samples
were treated in different ways, so that variable amounts of water
are incorporated in the analytical results. Therefore, it was necessary
to recalculate all analyses on a water-free basis.

In some cases, for elements which had not been determined by a given
laboratory, mean values from data from other laboratories had to be
used in the calculation. Those values are given in brackets in the
tables. In no case was it evident whether a sample "as received"
was analyzed, or some size fraction. This might account for some of the
observed discrepancies in the results, since some impurities are enriched
in the large grain size fractions.

OECD X-RAY FLUORESCENCE ANALYSIS

01 Montmorillonite

	F15	F19	D2	C.A.
SiO_2	64.00	64.45	64.68	63.84
Al_2O_3	22.99	23.23	22.46	22.24
TiO_2	0.56	0.58	0.49	0.50
Fe_2O_3	3.29	3.27	3.69	3.45
MnO				
MgO	5.86	(4.67)	4.69	4.87
CaO	3.14	3.27	3.49	2.88
Na_2O	(0.12)	(0.12)	0.12	1.73
K_2O	0.05	0.41	0.36	0.40

02 Laponite

	F15	C.A.
SiO_2	62.93	66.03
Al_2O_3	0.36	0.30
TiO_2	0.02	0.02
Fe_2O_3	0.04	0.06
MnO		
MgO	31.90	29.03
CaO	0.30	0.34
Na_2O		
K_2O	<0.01	0.04

03 Kaolinite (China Clay)

	F15	F19	D2	D20	C.A
SiO_2	52.31	51.60	53.60	51.76	53.53
Al_2O_3	43.58	45.20	44.00	44.40	43.95
TiO_2	0.04		0.04	0.07	0.03
Fe_2O_3	0.45	0.45	0.61	0.85	0.58
MnO					
MgO	0.25		0.20	0.08	0.17
CaO	0.09	0.80	0.10	0.09	0.20
Na_2O			0.09		0.08
K_2O	1.46	1.60	1.37	1.39	1.37

C.A.: Average data reported under "Chemical Analysis"

04 Attapulgite

	F15	F19	C.A.
SiO_2	76.20	74.45	75.15
Al_2O_3	8.61	9.67	9.72
TiO_2	0.61	0.70	0.74
Fe_2O_3	3.18	3.15	3.08
MnO	0.15	0.15	0.10
MgO	8.78	(8.74)	8.35
CaO	1.86	2.33	2.03
Na_2O			
K_2O	0.76	0.82	0.74

05 Illite

	F15	F19	D2	D20	C.A.
SiO_2	54.48	55.48	55.50	57.18	54.58
Al_2O_3	22.49	21.28	22.20	18.96	21.93
TiO_2	0.84	0.80	0.75	0.88	0.90
Fe_2O_3	7.62	7.21	7.40	8.21	7.41
MnO		0.05			0.03
MgO	4.10	(4.00)	3.87	3.14	3.76
CaO	1.36	1.37	1.28	1.42	1.25
Na_2O	(0.55)	(0.55)	0.55	(0.55)	0.39
K_2O	8.57	9.27	8.70	9.66	8.88

06 Chrysotile

	D16	C.A.
SiO_2	41.40	41.40
Al_2O_3	<0.3	0.73
TiO_2	<0.01	0.03
Fe_2O_3	2.04	1.84
MnO	0.06	0.04
MgO	41.80	42.30

C.A.: Average data reported under "Chemical Analysis"

OECD X-RAY FLUORESCENCE ANALYSIS

	08 Talc		11 Magnesite		
	D16	C.A.	D2	GB25	C.A.
SiO_2	60.00	60.36	6.00	4.04	5.43
Al_2O_3	0.80	0.70	1.11		0.98
TiO_2	0.05	0.07		0.026	0.06
Fe_2O_3	0.86	0.95	1.54	1.55	1.44
MnO	0.01	0.03		0.06	0.05
MgO	31.80	31.41	40.86	43.05	42.50
CaO			2.39	2.16	2.61
Na_2O			)<0.5		0.13
K_2O			)		0.09

Trace elements ppm

	01°	01	02	03	05	11	13
	D5	D 10	D5	D5	D 10	GB25	B3
Ba					349	670	
Co	43		~5	10	132	3	
Cr	~15						
Cu				~50		17	
Mn		62			410		
Ni	33		23	29		7	
Pb	9		5			11	
Rb				193	515		
Sn				20			
Sr	254	715	18	195	646	96	2360
V		139			168	3	
Zn							

(°) fraction <2 µm

C.A.: Average data reported under
"Chemical Analysis".

01 Montmorillonite

There is a main discrepancy between the values reported for MgO and for K_2O : F15 and D2 find respectively 5.86% and 4.69% for MgO; F15, F19, and D2 find respectively 0.05%, 0.41%, and 0.36% for K_2O.

Two laboratories determined a number of trace elements, however D5 used a 52 mesh sample and D10 a fraction <10 μm. Therefore, the results can not be compared.

03 Kaolinite (China Clay)

The analysis by D2 was recalculated assuming an ignition loss of 12.80%. There is fair agreement between laboratories, except for a high CaO value from F19.

04 Attapulgite

The two analyses by F15 and F19 differ mainly in the Al_2O_3 and CaO data.

05 Illite

There is fair agreement between the results from F15, F19, and D2, whereas D20 reports a higher SiO_2 and a much lower Al_2O_3 value. This discrepancy may be due to the fact that D20 did not fuse the sample. There is a greater difference between the K_2O values which range from 8.57% to 9.66%

11 Magnesite

D2 reports lower values for CO_2 and MgO than GB25, indicating a lower magnesium carbonate content, whereas the SiO_2 value of D2 is 2% (absolute) higher. Perhaps, the analyzed samples were not identical.

DISSOLUTION METHODS

U. Schwertmann

INTRODUCTION

(1) <u>General</u>

Selective dissolution methods (see also Wada and Harward),
1974, Dixon and Weed, 1977), as applied to clay-containing
samples have two main purposes:

(a) to remove certain contaminants and thereby to facilitate
further study of the main components (e.g. formula computation);

(b) to determine quantitatively in natural mixtures of
secondary (and primary) minerals (sediments, soils, etc.),
compounds which are difficult, if not impossible, to estimate
otherwise because of their non-crystalline nature or because
of the small amounts present.

The latter is perhaps the most important application.

To fulfill these purposes, dissolution methods should be
as selective as possible. The selectivity is improved by
choosing conditions of extraction under which the solubilities
of the various components differ greatly. This approach is,
however, not always possible and difficulties in interpretation
then arise. Furthermore, the methods should be reasonably
rapid and simple.

It must be recognized that each technique has its advantages and disadvantages, and that an intelligent selection must therefore be made to suit the problem at hand.

Selective dissolution methods are widely used for the extraction of soluble salts, carbonates, sulphides, oxides and hydroxides of Si, Ti, Al, Fe, Mn, and organic compounds. However, they have been extended to determine the clay silicates themselves (e.g. kaolinite after dehydroxylation by heating the sample to 525 ^{o}C), and, indirectly, to determine quartz and feldspars after the layer silicates have been removed by pyro-sulphate fusion. (Alexiades and Jackson, 1966).

(2) Methods

Soluble salts, including gypsum

These compounds may be easily extracted with water. The sample-water ratio is determined by the solubility of the salts present. With clay samples, difficulties may arise due to dispersion of the clay, especially in the presence of sodium. In this case, soluble salts and exchangeable cations have to be extracted together using an electrolyte solution in which the extracted ions are then equivalent to the soluble salts.

Carbonates

If the carbonates are to be removed for cleaning purposes, only a very mild extractant should be used. The use of dilute HCl with permanent pH control (not below pH 3.5 - 4.0) or a 1N sodium acetate solution adjusted to pH 5 is recommended. (Alexiades and Jackson, 1966). For a quantitative determination of total carbonates, the volumetric measurement of the CO_2 evolved on HCl addition is widely used, as well as a conductometric and titrimetric determination of CO_2. A separate determin-

ation of calcite and dolomite (if not performed by X-ray diffraction or by a microscopic staining method) is also possible by
a dissolution method. This determination is based on the difference in rates of solution of these two minerals. Solution vs.
time curves exhibit a non-linear, followed by a linear part, which
on extrapolation to zero gives the dolomite content (see Skinner,
Halstead, and Brydon, 1959).

Sulphides

On treating the sample in an N_2 or a CO_2 atmosphere with concentrated HCl (d = 1.16), H_2S is evolved, which is collected
in an acetic acid solution containing Cd and Zn acetate, and
titrated iodometrically (Purokoski, 1958 ; Rasmussen, 1961).

Oxides and hydroxides of Si, Al, Fe, and Mn.

Silicon and Aluminum

Alkaline solutions are used for the extraction of "free silica"
and "free alumina". The main problem is that there is no sharp
difference in solubility between non-crystalline components
(silica and alumino silicates) and fine grained crystalline
silicates.

The method proposed by Foster (4 hours extraction with
0.5 N NaOH at $100^\circ C$) appears to be too drastic because considerable amounts of crystalline silicates are brought into
solution. A milder extraction is applied in the procedure
proposed by Hashimoto and Jackson (1960), who reduced the
extraction time applied in the Foster method to 2.5 minutes,
and that by Follett et al. (1965) who proposed successive
extractions with 5 % Na_2CO_3 solution at room temperature,
or on a steam bath.

Using these procedures, quartz and well crystallized clay silicates are generally unaffected, whereas non-crystalline silica, aluminum hydroxide, and allophane are dissolved (see Duchaufour et al., 1966). Gibbsite will be dissolved by the Hashimoto and Jackson procedure but not by the Follett et al. procedure. However, crystalline gibbsite can be determined better by X-ray diffraction or DTA after removing those crystalline iron oxides which are likely to interfere.

Halloysite amd also smectite, particularly when of a small particle size (a few tenths of one micron) will be dissolved during the NaOH treatment after Hashimoto and Jackson (Askenasy et al., 1973; Higashi and Ikeda, 1974; Wilke et al., 1977). Therefore, a different treatment was developed employing acid ammonium oxalate at room temperature (Fey and Le Roux, 1977; Kitagawa, 1976), at $30^{\circ}C$ (Higashi amd Ikeda, 1974) or at $90^{\circ}C$ (Henmi and Wada, 1976). Further methods proposed for the quantitative determination of allophane are based on alternating treatment with 8 N HCl and 0.5 N NaOH (Kitagawa, 1976), on cation exchange capacities delta value (Aomine and Jackson, 1959), on differential infrared absorption spectroscopy (Wada and Tokashiki, 1972) and on weight loss between 105 $^{\circ}C$ and $200^{\circ}C$ (Kitagawa, 1976).

Iron

The removal of so-called "free iron oxide" prior to the X-ray diffraction examination of clays may be useful since iron oxides tend to aggregate clay particles, thereby preventing proper orientation and increasing the background count. Therefore, X-ray patterns are usually improved by such treatment.

This removal is readily performed with a reducing agent.

Sodium dithionate combined with $NaHCO_3$ as a buffer and Na-citrate
as a complexing agent for the iron (as proposed by Mehra and
Jackson, 1960) is widely used for this purpose. Silicates as well
as allophane are only slightly attacked, with the exception of
biotite in which some removal of potassium occurs. Samples
containing greater amounts of iron oxides, concretions or well-
crystallized oxides usually require more than one extraction.
Considerable amounts of Al are found in the dithionate extract,
which originates either from isomorphous Al substitutions in
goethites or from "free alumina" extracted by the citrate in the
extractant. Therefore, the dithionate-citrate method appears to be
a method by which both Fe and Al are extracted but also some Si.
An alternative procedure has been proposed by Duchaufour and
Souchier (1966), who employed dithionate plus oxalate.

Iron oxides of any crystallinity are removed by the dithionate
method. A method for the separation of very poorly crystalline
iron oxides, particularly ferrihydrite (formerly called amorphous
ferric hydroxide) from better crystallized species (especially
goethite and hematite) has been developed by Schwertmann (1964)
using an NH_4-oxalate/oxalic acid mixture at pH 3.0, similar to
the TAMM solution (see also McKeague and Day, 1966). During the
extraction, sunlight must be excluded to inhibit photochemical
reduction which renders crystalline oxides soluble.

This separation method has been checked by X-ray, thermal,
and EM techniques to be reasonably reliable for natural and
synthetic samples (Gast, Landa, and Meyer, 1974). For these
samples the oxalate soluble proportion varies between less
than 0.1 % and 100 %, and is related to aging conditions of
the iron oxides. Lepidocrocite, in contrast to goethite and
hematite is slightly soluble in oxalate, particularly when of

a very small crystal size (Pawluk, 1972; Schwertmann, 1973;
Schwertmann and Fitzpatrick, 1977). Also, organically bound
as well as exchangeable iron is extracted, but these usually
constitute only a small proportion. Some Fe from freshly
ground chlorites and biotites can also be dissolved during
the oxalate extraction (Arshad et al., 1972). An alternative
procedure to remove poorly crystalline Fe-oxides has been
developed by Segalen (1968) employing eight consecutive
alternating extractions with 8 N HCl and 0.5 N NaOH and
extrapolating the resulting extraction curves.

Manganese

Secondary manganese oxides are almost completely dissolved by the dithionate-citrate treatment (as checked
with pyrolusite, and with sediments and soils) so that a
Mn determination of this extract is recommended. Similar
amounts of Mn occur in the acid oxalate extract (Blume
and Schwertmann, 1969). Extraction with dithionate or oxalate
does not separate Mn from Fe oxides. This separation can
be achieved using H_2O_2 in acid solution as proposed by
Taylor and McKenzie (1966). Strong acids, on the other
hand, bring the Mn of crystalline silicates into solution
(as checkd with illite and with biotite), and are, therefore,
not suitable.

Organic carbon

A hydrogen peroxide treatment is recommended for removing
organic matter, although this treatment is not always success-
ful. For a rapid quantitative determination of organic carbon
wet oxidation by dichromate in concentrated H_2SO_4, followed
by photometric determination of the Cr^{3+} concentration at

578 nm is used. Alternatively, CO_2 can be determined
titrimetrically. In this case, CO_2 from carbonate must be
determined separately.

Other Methods

Kaolinite minerals

Hashimoto and Jackson (1960) proposed that kaolinite
minerals be determined quantitatively from the amounts of
Si and Al brought into solution by boiling for 2.5 minutes
with 0.5 N KOH after dehydroxylation the K-saturated sample
at 525 $^{\circ}$C. The amounts of Si and Al dissolved when using an
oven dried sample (110 $^{\circ}$C) must be subtracted. Some fine-
grained 2:1-layer silicates are also dissolved in both
treatments. In kaolinite-gibbsite samples the kaolinite
and gibbsite content may be determined from the Si and Al
content of a single extract after heating to 525 $^{\circ}$C. The
kaolinite content is then calculated from the Si content and
the gibbsite content from the Al not attributed to the
kaolinite (LeLong,1967).

Kaolin minerals can also be removed, e.g. in order to
concentrate iron oxides, by boiling the sample for 0.5 to
1 hour in 5 N NaOH and removing the sodalite formed by a
subsequent HCl wash (Norrish and Taylor, 1962).

Quartz and feldspars

In a sample containing layer silicates, feldspars and
quartz, the last two are found in the residue after fusion
with $Na_2S_2O_4$. The feldspars can then be determined from the
K, Na, and Ca content of the residue (Kiely and Jackson,
1964).

Rutile and anatase

The sum of these two minerals is determined in the residue
from an HF treatment by dissolving this residue in H_2SO_4-
$HClO_4$-HF mixture and determining its Ti content photometrically

with iron (Raman and Jackson, 1965). Recently Dolcater et al. (1970) and Sayin and Jackson (1975) concentrated the free TiO_2 phases in kaolins by dissolving kaolinite in hexafluorotitanic acid (H_2TiF_6). Amorphous and very poorly crystalline Ti-oxides can be successfully separated from crystalline species using the acid ammonium oxalate method (Fitzpatrick et al., 1977).

References

Alexiades, C.A. and Jackson, M.L. (1966), Clays and Clay Minerals, 14, 35–68

Aomine, S. and Jackson, M.L. (1959), Soil Sci.Soc.Amer.Proc., 23, 210–214

Arshad,M.A., St.Arnaud, R.J. and Huang, P.M. (1972), Can.J. Soil Sci., 52, 19–26

Askenasy, P.E., Dixon, J.B. and McKee, T.R. (1973), Soil Sci. Soc.Amer.Proc., 37, 799–803

Blume, H.P. and Schwertmann, U. (1969), Soil Sci.Soc.Amer.Proc., 33, 438–444

Dixon, J.B. and Weed, S.B., Editors (1977), Amer.Soc.Agron., Madison, Wis., U.S.A.

Dolcater, D.L., Syers, J.K. and Jackson, M.L. (1970), Clays and Clay Minerals, 18, 71–79

Duchaufour, Ph. and Souchier, B. (1966), Sci. du Sol, 17–30

Fey, M.V. and Le Roux, J. (1977), Clays and Clay Minerals, 25, 285–301

Fitzpatrick, R.W., Le Roux, J. and Schwertmann, U. (1977), Clays and Clay Minerals, 25, 373–374

Follett, E.A.C. et al. (1965), Clay Minerals, 23-43

Foster, M.D. (1953), Geochim. et Cosmochim. Acta, $\underline{3}$, 143-154

Gast, R.G., Landa, E.R. and Meyer, G.W. (1974), Clays and
Clay Minerals, $\underline{22}$, 31-39

Hashimoto, I. and Jackson, M.L. (1960), Clays and Clay Minerals,
$\underline{7}$, 102-113

Henmi, T. and Wada, K. (1976), American Mineralogist,
$\underline{61}$, 379-390

Higashi, T. and Ikeda, H. (1974), Clay Science, $\underline{4}$, 205-211

Kiely, P.V. and Jackson, M.L. (1964), American Mineralogist,
$\underline{49}$, 1648-1659

Kitagawa, Y. (1976/7), Soil Sci. Plant Nutr., $\underline{22}$, 137-147;
$\underline{23}$, 21-31

LeLong, F. (1967), Bulletin Groupe Français des Argiles,
$\underline{19}$, 48-67

McKeague, J.A. and Day, J.H. (1966), Can. J. Soil Sci,
$\underline{46}$, 13-22

Mehra, O.P. and Jackson, M.L. (1960), Clays and Clay Minerals,
$\underline{7}$, 317-327

Norrish, K. and Taylor, R.M. (1962), Clay Minerals Bulletin,
$\underline{5}$, $\underline{28}$, 98-109

Pawluk, S (1972), Can. J. Soil Sci., $\underline{52}$, 119-123

Purokoski, P. (1958), Agr. Publ. of Finland,No. 7a

Raman, K.V. and Jackson, M.L. (1965), American Mineralogist,
$\underline{50}$, 1086-1092

Rasmussen, K. (1961), Transformations of inorganic sulphur
compounds in soil. Copenhagen, 176 pp.

Sayin, M. and Jackson, M.L. (1975), Clays and Clay Minerals, 23, 437-443

Schwertmann, U. (1964), Z. Pflanzenern., Dgg., Bodenk., 105, 194-202

Schwertmann, U. (1973), Can. J. Soil Sci., 53, 244-246

Schwertmann, U. and Fitzpatrick, R.W. (1977), Soil Sci. Soc. Amer. Proc., 41, 1013-1018

Segalen, P. (1968), Cah. O.R.S.T.O.M. Ser. Pedol., 6, 105-126

Skinner, S.I.M. et al. (1959), Can. J. Soil Sci., 39, 197-204

Taylor, R.M. and McKenzie, R.M. (1966), Aust. J. Soil Res., 4, 29-39

Wada, K. and Harward, M.E. (1974), Advanc. in Agron., 26, 211-260

Wada, K. and Tokashiki, Y. (1972), Geoderma, 7, 199-213

Wilke, B.-M. et al. (1977), Clay Minerals, 13, 67-77

OECD RESULTS

Only a small number of determinations were carried out. These
included methods for determining the amount of amorphous oxides and
hydroxides of Si, Al, and Fe, and methods for the quantitative
determination of crystalline compounds which are difficult to
determine by other methods.

The results obtained by the three participating laboratories
on the content of amorphous impurities are shown in the table.
One laboratory determined crystalline compounds in the China
Clay by two different methods (GB5). The results of these determinations are: mica: 11-13%; quartz: 0.5-1.5%; kaolinite: 85-90%.

OECD COMMENTS

(a) Methods for removal and quantitative determination of amorphous
contaminants.

01 Montmorillonite

The montmorillonite sample contains about 2% alkali soluble
silica (cryptocrystalline silica or fine quartz) and 0.4% soluble
Al oxides.Therefore, it is also quite low in amorphous impurities.
The content of free iron is at most 0.2%.

05 Illite

The illite sample contains about 2% alkali soluble silica
(cryptocrystalline silica or fine quartz), about 1% soluble
Al oxides, and less than 0.1% free iron oxides, hence it is
extremely low in amorphous constituents. Dithionite-bicarbonate
does not remove as much Si and Al as is achieved in the alkaline
methods, due to its lower extraction pH.

10 Gibbsite

Gibbsite, which is well crystallized according to its X-ray diffraction pattern is only slightly soluble in hot 5% Na_2CO_3 (9.3%
Al_2O_3), but considerably,although not completely soluble in boiling
0.5 N NaOH (46.8% Al_2O_3, equivalent to 71.4% $Al(OH)_3$). The authors
of this dissolution method assumed that gibbsite would be insoluble
in hot Na_2CO_3, and completely soluble in boiling NaOH. The different
results mentioned here may be due to particle size effects, as well
as to the reaction times applied. The solubility of gibbsite in
$Na_2S_2O_4$+ $NaHCO_3$ is very low, it may be increased by the addition of
citrate.

The content of alkali soluble silica as well as reducible iron
is extremely low, indicating that the sample is rather pure, containing only a few tenths of one percent of SiO_2 and less than
0.1% Fe_2O_3.

OECD: ANALYSIS BY DISSOLUTION METHODS

Lab.	SiO_2				Al_2O_3				Fe_2O_3		
Method(°)	GB15 Na_2CO_3	D6 NaOH	GR4 NaOH	GB15 $Na_2S_2O_4$	GB15 Na_2CO_3	D6 NaOH	GR4 NaOH	GB15 $Na_2S_2O_4$	GB15 $Na_2S_2O_4$	GR4 $Na_2S_2O_4$	D6 $Na_2S_2O_4$
01 Montmor.	2.0	2.26	8.3	0.2	0.4	0.42	1.3	0	0.2	0.61	0.07
05 Illite	2.1	2.14	n.d.	0.3	0.7	0.23	n.d.	0.1	0.1	n.d.	0.03
10 Gibbsite	0	0.3	n.d.	0.1	9.3	46.8	n.d.	0.9	0.04	n.d.	0

(°) Na_2CO_3 : one cold and three hot 5% Na_2CO_3 treatments after Follett et al

    NaOH : boiling for 2.5 minutes in 0.5 N NaOH after Hashimoto and Jackson

    $Na_2S_2O_4$: Na-dithionite buffered with bicarbonate without (GB15) and with (GR4 and D6) citrate.

(b) <u>Methods for the quantitative determination of crystalline</u>
    <u>compounds</u>

Laboratory GB5 used two different methods to determine the
kaolinite content of China Clay, while at the same time isolating
other materials.

1. The first method is the method by Hashimoto and Jackson, applying
2.5 minutes boiling of 3 g in 0.5 N NaOH after dehydroxylation of the
sample by heating for 24 hours at 500°C. Non-kaolinite residue amoun-
ted to 14%. According to X-ray analysis this residue consists of about
50% mica and 40% mixed-layer minerals. The latter are of two kinds, one
showing a 001 spacing of 27.6 A after glycerol treatment, the other showing
a 001 spacing of 23.8A with two or three higher orders indicating the
formation of respectively mica-montmorillonite and mica-vermiculite.
These mixed layer components probably resulted from the attack of NaOH
on the mica present. Some of the kaolinite was still present after the
treatment. The total composition of the sample derived from these data
is: 12-13% mica, 0.5-1.0% quartz, and the remainder kaolinite.

2. The second method (by Malden, unpublished) consists of boiling
20 g of the sample for 2 hours in 2.5 N NaOH after 16 hours of heating
at 500°C, and grinding the heated sample with 2.5 N NaOH as a paste.
After boiling, the sample was washed with water and HCl at pH 2.0
for 25 minutes. The results obtained were very similar to those
of the first method, i.e. 11% mica, 0.5% quartz, the remainder
kaolinite. Less mica was converted to mixed layer minerals and
only mica-vermiculite was formed. The method is less cumbersome than
the first method, yielding sufficient residu for analysis since
20 g rather than 3g is used.

<u>Spread of the data</u>

In view of the small number of data, little can be said about
the observed spread of the results from different laboratories.
In addition, the applied methods are not very well defined,
therefore, slight deviations in the details of the procedures

may occur, and these have relatively large effects with samples containing small amounts of amorphous constituents;

It is encouraging that for the illite and montmorillonite samples two of the three laboratories obtained similar results for silica and alumina, applying the NaOH and Na_2CO_3 methods. The third laboratory obtained much higher results and this could well be due to a somewhat longer boiling period which should be exactly 2 minutes.

For the determination of amorphous Al oxides in gibbsite, a mild extraction procedure is recommended, i.e. treatment with Na_2CO_3 in the cold in order not to dissolve gibbsite.

The spread in the data for dithionite soluble iron is surprisingly large. No explanation of the causes of these discrepancies can be given.

X-RAY DIFFRACTION[+]

K. Jasmund and J. Mering

INTRODUCTION

The following procedural suggestions apply to a qualitative
phase analysis by X-ray diffraction.Procedures for a quantitative
analysis would be somewhat different.Only a general indicative guide
is given.

Particle size grading

Samples containing large particles must be triturated for
diffraction examination. For a fixed specimen holder the
particle size must be reduced to sizes <10 μm diameter, whereas
for moving (e.g. rotating) specimen holders <30 μm diameter
would possibly suffice. If grinding is done in a ball mill, the
crushing of the powder is best carried out under a liquid (water,
cyclohexane, etc.). Care must be taken with grinding particularly of
layer minerals,because the crystalsmay suffer deformation, especially
during prolonged grinding. Therefore, hand-grinding with pestle and
mortar is preferred with repeatedly removing the fine fraction so that
grinding is always done on the larger particles. If the material itself
is fine-grained, a particle size 100 m is acceptable so that less
disorder is introduced by grinding.

Sample preparation

Special precautionary measures are necessary with minerals
of easy cleavage, and particularly with layer minerals. In sample

---

(+) X-ray diffraction results were reported only for the OECD suite
of samples.

preparation for such minerals, two types of specimen must be
considered:

(a) largely· parallel orientation of particles on their basal
surfaces ("oriented aggregates");

(b) largely   random orientation of particles ("unoriented
specimens").

Between these two extremes, various degrees of orientation are
possible, but for such specimens interpretation of the results
is very difficult. Parallel orientation assists in assigning a
layer silicate to a specific group by enhancing its basal reflections
with respect to the reflections of non-platy minerals or minerals
with less perfect cleavage.. If useful relative intensity measurements are to be made, however, the specimen should be
unoriented, since any orientation will affect the intensity ratio
of basal to non-basal reflections. In unoriented specimens, the
intensities, as well as the positions of the lines can be used
in the characterization of the sample. However, the identification
using intensities will be complicated if the reference mineral
occurs in mixtures with other phases, especially when diffraction
patterns have a large number of reflections.

(1) Sample preparation for the X-ray diffractometer (Bragg-Brentano)

(a) Unoriented specimen

For the preparation of specimens without orientation the method
of Niskanen (1964) is recommended. For filling the specimen holder
a roughened glass plate is used to prevent orientation of the flakes
such as would occur if they were arranged in a parallel fashion on
a smooth glass surface. This method is also well suited for quantitative determinations on mixtures.

(b) Oriented aggregates, i.e. parallel orientation of flakes.

A suitable technique is described by Kinter and Diamond (1956)
This technique is particularly useful for the detection of small
amounts of clay minerals mixed with the main component in a clay
sample. Deposition of particles on a porous tile by suction is also suitable.

(2) Sample preparation for powder cameras.

(a) Oriented aggregates.

Oriented aggregates can be used in a Debye-Scherrer camera
(Brown, 1953; Mitchell, 1953), or in a Brindley-Robinson camera (°)

° The Brindley-Robinson camera is equally suited for random samples,
but its use is superceded by the diffractometer.

(Perrin, 1955), or in a camera using curved specimens (Jasmund, 1956a).The latter has the advantage that short exposure times can be used. Different methods of sample preparation, e.g. by sedimentation, centrifuging,deposit on ultrafilters) will, however, lead to different degrees of orientation.

With oriented aggregates usually only the basal spacing peaks are recorded, but the non-basal spacing peaks can also be registered using a Guinier camera. One of the advantages of this method is that only very small amounts of substance are necessary for the examination.

(b) Unoriented specimens

A minimum degree of orientation can be obtained in different ways, for example by pressing a paste between two cellophane sheets (Jasmund,1956b). Another very simple method is to stir the powder with a viscous varnish, e.g. cellulose varnish and allowing it to dry to a thin skin. For other methods see Flörke and Saalfeld, 1955; Brindley and Kurtossy, 1961).

A special pretreatment is advisable for swelling clay minerals. Such samples should be saturated with Mg or Ca ions prior to examination, and the humidity during the determination should be controlled and recorded.

Interpretation and instrument conditions

The intensity ratio of characteristic lines, e.g. first and second order basal lines and 060 lines of clay minerals, should be calculated by integral planimetering or integral counting statistics after making the necessary background corrections. From these data one can estimate the degree of orientation in the specimen (Brindley and Kurtossy, 1961). A check by other methods is also desirable. Statements on the degree of structural order should be made using the half-peak width ratios of appropriate lines. In reporting results, instrument conditions should be described

Identification of Clay Minerals

With clay minerals, identification is not always immediately obvious. X-ray diffraction is actually only one way to describe their properties, and this information must sometimes be combined with information obtained by other techniques to allow a proper identification. The following are some of the problems which are encountered in the identification of clay minerals by X-ray diffraction data.

(a) It is not always easy to distinguish a smectite from a
vermiculite in the whole range of hydrated triform minerals.
It is recommended to saturate the mineral with potassium ions.
Potassium vermiculite, heated for several hours to 50-100°C
undergoes irreversible dehydration which is detectible by X-ray
methods. This phenomenon does not occur with smectites.

(b) In the smectite group, it is sometimes difficult to distinguish a montmorillonite from a beidellite (dioctahedrial mineral)
or a hectorite from a saponite (trioctahedrial mineral). When the
diffraction pattern reveals an ordered or a semi-ordered structure
it may be assumed that the clay is a beidellite or saponite
respectively. On the other hand, the absence of order by itself is
insufficient evidence that no beidellite or saponite is present.

To distinguish a montmorillonite from a beidellite, it is recommended to apply the Hofmann-Klemen test: saturate the sample with
Li, heat for several hours at 300°C, immerse for several hours
in ethylene glycol and examine the treated mineral by X-ray
diffraction. Formation of the normal clay-glycol complex indicates
a beidellite. By measuring the cation exchange capacity of the
complex after removal of glycol, it is possible to identify a
mixed mineral of the montmorillonite-beidellite series.

Unfortunately this method is not applicable to tri-octahedric
minerals, but for those the results of a chemical analysis for
Li and Al will be a good guide.

(c) Clay minerals, or in general minerals with lamellar structures
often occur as interstratified minerals. In order to identify and
describe interstratification,observation of only the first basal
reflection should never be regarded as adequate. It is essential
to measure the complete series of these reflections; for a disordered
sequence, the concept of Bragg reflection and integral orders breaks
down.

References

Brown,G., "X-ray identification and crystal structures of
    clay minerals". Mineralogical Society, London, 1961

Brown,G.(1953) J.Soil Sci., $\underline{4}$, 229

Brindley,G.W.and Kurtossy,S.S. (1961) Amer. Min. $\underline{46}$, 1205

Flörke,O. and Saalfeld,H. (1955) Z.Krist.,$\underline{106}$, 460

Jasmund,K. (1956a) Naturw., 43, 275

Jasmund,K. (1956b) Geol.Fören. Stockholm Förh., 78, 156

Kinter,E.B. and Diamond,S. (1956) Soil Sci., 81, 156

Mitchell, W.A. (1953) Clay Min.Bull., 2, 76

Niskanen, E. (1964) Amer.Min., 49, 705

Perrin,R.M.S. (1955) Clay Min.Bull., 2, 307

OECD RESULTS

For most of the samples a large number of laboratories reported X-ray diffraction data, but patterns varied considerably in quality and detail. The majority of the laboratories used a diffractometer with rate meter, i.e. a comparatively rapid, but not very precise method. Generally, a scanning rate of 0.5°(2θ) was used.

The following tables present selected spectra for each sample in the form of a list of d-values and intensity measurements as reported by the authors, i.e. without identifying the precision of the data. It should be noted that without extraordinary care, intensities can only be given on a scale of broad intervals : between 100 and 30 intervals of 5, between 30 and 16 intervals of 2 and between 15 and 0 intervals of 1 . Sample pretreatment, if any, and conditions of measurement are indicated. In each table a reference spectrum of the principal mineral is included, some with calculated d-values. These were taken from Brown (l.c.), or from the X-ray Powder Diffraction File (X.P.D.F.). For assignments of observed peaks the referenced literature should be consulted.

For the clay minerals 01-05 the full reported spectrum is shown, including peaks belonging to impurities. For the other minerals,peaks belonging to impurities are listed separately.

01   Montmorillonite

lit.°	B1°°		B1°°°		F6°°		F6°°°		JAP3		B1 Glycol	
calc.	d(A)	I	d(A)	I	d(A)	I	d(A)	I	d(A)	I	d(A)	I
	11.95	41	12.63	94	14.7	10	16.4	10	15.0	100	17.16	53
			6.24	6					4.96	11	10.05	2(M)
4.50	4.48	56			4.48	4	4.48	6	4.49	20	8.55	4
									4.25	34 Q	5.68	4
					4.15	2	4.11	3			4.30	1
					3.92	2					3.38	4(M,Q)
					3.67	2	3.76	3	3.70	9	3.21	2(F)
					3.51	2					2.83	1
			3.34	4	3.36	2	3.38	4	3.34	2 M,Q		
			3.21	6	3.22	3	3.22	7	3.187	2 F		
	3.14	28			3.15	3						
			3.11	9								
							3.05	4	3.028	4. C		
							2.95	2	2.969	9		
					2.75	2			2.823	2		
2.60	2.57	20			2.54	4	2.54	4	2.536	34		
									2.356	8		
2.25												
					2.12	2	2.12	2				
					1.83	1	1.83	2				
1.706	1.68	7			1.68	3	1.68	2	1.695	6		
									1.675	5		
1.503	1.50	16					1.49	3	1.498	12		
1.301	1.29	7							1.296	2		
									1.290	3		
1.252	1.25	5							1.247	4		

(°) Brown l.c. p. 192-193 (°°) disoriented (°°°) oriented
B1 : Ca form, fraction <10µ m ; JAP3 :fraction <37µ m
B1 : CuKα ,30 kV,20 mA; F6 : CuKα ,37 kV, 7 mA ;
  JAP3 : CuKα ,40 kV, 20 mA
Impurities :
    M: Mica; Q: Quartz; F: Feldspar; C: Calcite
      2-3%   3-5%   6-10%   1-2%
B1°°, R.H. 45%; B1°°°, R.H. 57%

02   Laponite

lit.°	B1°°		B1°°°		D14		JAP3		B1 Glycol	
calc.	d(A)	I	d(A)	I	d(A)	I	d(A)	I	d(A)	I
	14.37	27	13.49	56	14.0	30	15.8	90	17.67	69
			5.91	4			5.34	18	9.03	9
4.50	4.53	17			4.48	20	4.53	44	5.68	5
	3.21	8	3.16	11			3.08	9	4.33	3
2.60	2.56	12			2.580	10B	2.57	100	3.40	10
2.25									2.81	3
1.706	1.71	3					1.725	10		
1.503	1.52	10			1.520	6B	1.523	24		
1.301	1.30	4					1.312	18		
1.252							1.264	5		

(°) Brown l.c. p. 192-193 (°°) disoriented (°°°) oriented
B1 : CuKα , 30 kV, 20 mA; D14: CuKα , 40 kV, 20 mA;
    JAP3 : CuKα , 40 kV, 20 mA
D14 ; fraction <20 µm; JAP3, fraction <37 µm.
B1°°, R.H. 42%; B1°°°, R.H. 46%; JAP3, R.H. 64%

03   Kaolinite (China Clay)

lit.°		B1°°		B1°°°		F6°°		F6°°°		F9		
d(A)	I	d(A)	I	d(A)	I	d(A)	I	d(A)	I	d(A)	I	
		10.05	7	10.05	4	10.00	2	10.00	3	10.04	8	M
7.16	10+	7.20	67	7.14	52	7.05	10	7.05	10	7.18	100	
		5.01	3	4.98	2	5.06	1	5.06	2	5.01	3	M
4.46	4	4.47	17					4.45	3	4.46	13	M
4.36	5	4.35	17	4.31	3	4.35	3	4.35	5	4.36	16	
4.18	5	4.17	15	4.16	3	4.19	2	4.19	4	4.17	12	
4.13	3									4.00	5	
3.845	4	3.87	9	3.85	2	3.84	1			3.86	5	M
3.741	2	3.74	6							3.74	5	M
3.573	10+	3.58	43	3.58	49	3.54	8	3.54	10	3.57	65	
3.372	4											
		3.33	8	3.33	7	3.33	2	3.33	5	3.34	7	M,Q
		3.21	4							3.21	2	M,F
3.144	3							3.12	1	3.12	1	
3.097	3					3.00	1	3.00	2	3.00	2	
		2.99	3				1	2.97	2			M
		2.87	2			2.86	1	2.86	1	2.86	2	M
2.753	3	2.80	2					2.77	1	2.79	2	
2.558	6	2.56	14	2.56	3	2.56	3	2.56	4	2.56	12	M
2.526	4	2.54	9			2.55	4	2.55	4			
2.491	8	2.50	12	2.49	3					2.50	10	
		2.46	2									M
2.379	6	2.38	6	2.38	6	2.38	2	2.38	6	2.38	7	
2.338	9	2.34	19	2.36	4	2.34	5	2.34	5	2.34	17	
2.288	8	2.29	11	2.29	2	2.29	4	2.29	3	2.29	10	
2.247	2											
2.186	3	2.20	3					2.19	1	2.19	2	
2.131	3	2.14	2					2.13	2			
2.061	2											
1.989	6	1.99	6	1.99	4	1.99	3	1.99	5	1.996	6	M
1.939	4	1.94	3					1.94	1	1.943	2	
1.896	3	1.90	2					1.90	1	1.901	1	
1.869	2											
1.839	4	1.84	2					1.83	1	1.847	2	
1.809	2											
1.781	4	1.79	2	1.79	3	1.78	1	1.78	4	1.789	3	
1.707	2											
1.685	2											
1.662	7	1.67	8	1.66	2	1.66	4	1.66	4	1.665	6	
1.619	6	1.62	4	1.62	1	1.61	2	1.61	2	1.622	2	
1.584	4	1.59	2	1.58	1							
1.542	5B	1.54	2	1.54	1	1.54	1	1.54	2	1.545	2	
1.489	8	1.49	9	1.49	2	1.48	4	1.48	3	1.489	7	M
1.467	2	1.46	2									
1.452	4B											
1.429	4			1.43	1	1.42	1	1.42	2			
1.403	2											
1.390	2											

5(°) Brown l.c. p.&111-112 (°°) disoriented (°°°) oriented
M : Muscovite mica, Q : Quartz ; F : Feldspar
D15 : Mica 9-10%; Quartz 1-2%; Feldspar trace.
D28 : fraction <2 µm Mica 3%; Quartz 1%.
GB5 : Mica 11%; Quartz 1%, fraction<<2 µm Mica 7%
B1 : CuKα , 30 kV, 20 mA; F6 : CuKα , 37 kV, 7 mA
B1°°, R.H. 45%; B1°°°, R.H. 42%

04 Attapulgite

calc. d(A)	lit.° d(A)	I	B1°° d(A)	I	B1°°° d(A)	I	F6°° d(A)	I	F6°°° d(A)	I	D12 d(A)	I	
10.48	10.50	10	10.46	71	10.57	77	10.7	10	10.7	7	10.5	vs.	
6.45	6.44	6	6.42	10	6.42	11			6.45	2	6.44	mw	
5.44	5.42	5	5.40	7	5.39	3			5.45	2	5.43	mw	
5.24													
					5.01	2					5.02	vw	M
4.50	4.49	8	4.46	17	4.47	6	4.49	3	4.48	3	4.48	m	
							4.33	4					
			4.27	10			4.28	3	4.29	5	4.26	ms	Q
4.18	4.18	3	4.14	10			4.18	3			4.15	mw	
			3.99	3							3.99	vw	F
			3.87	5	3.87	3							
3.69	3.69	5	3.65	5	3.66	5			3.70	1	3.66	mw	
3.49) 3.47)	3.50	3			3.40	6	3.43	5	3.38	10			
			3.34	10			3.34	4			3.33	s	Q,M
3.23	3.23	10	3.21	13	3.20	12	3.25	3			3.24	m	
											3.19	mw	F
									3.12	1	3.10	mw	F
3.04	3.03	3	3.04	59	3.04	22			3.03+	2	3.03	w	C
											2.99	w	M
2.76											2.89	vw	D
			2.67	3							2.68	w	F
2.62	2.61	8	2.59	9	2.60	4	2.61	1	2.63	1	2.63	mw	
											2.59	mw	
2.56	2.55	3	2.54	12			2.56	1	2.55	3	2.53	mw	
			2.50	13	2.50	3	2.47	1	2.47	3	2.51	mw	F
											2.45	m	Q
2.38	2.38	3									2.37	w	
			2.29	9	2.29	2	2.29	1	2.29	2	2.27	mw	Q
2.25									2.24	1	2.23	mw	Q
											2.21	vw	
2.15	2.15	5	2.17	3							2.16	w	
			2.12	5	2.12	2			2.13	2	2.12	mw	Q
2.10			2.10	8			2.09	3	2.10	1			
									1.98	2			Q
			1.91	8	1.91	2							C
			1.88	8	1.87	2							C
1.845													
1.815	1.82	1					1.82	2	1.82	4	1.82	m	Q
1.75													
			1.68	3					1.67	2	1.67	m	Q
1.615	1.62	1	1.61	4							1.61	w	Q
1.555	1.56	3	1.53	5					1.54	3	1.54	m	Q
1.50	1.50	5	1.51	5									

(°) Brown l.c. p.352   (°°) disoriented   (°°°) oriented

B1 : CuKα   , 30 kV, 20 mA; fraction   <10 μm.

F6 : CuKα   , 37 kV, 7 mA

D12: CuKα   , 30 kV, 20 mA

(+) Peak disappears after decarbonation.

Impurities/ M:Mica; Q:Quartz; D:Dolomite; C:Calcite; F:Feldspar

Determinations of quartz: GB5 : 24%; D12 : 22% (using CaF_2 as internal standard)

  F13 : 26% ± 1.5% (Bull.de la Soc.Française de Céramique, 38, 29, 1958)

B1°°, R.H. 33%; B1°°°, R.H. 54%

05 Illite

St.Austell°		F6°°		F6°°°		D28 +	
d(A)	I	d(A)	I	d(A)	I	d(A)	I
10.1	s	10.03	10	10.02	10	10.1	10
4.98	m			5.07	1	5.00	2
4.50	s	4.52	10	4.52	7	4.49	6
4.35	vw					4.32	<1
				4.27	3		
4.10	vw						
3.85	vw						
		3.78	4	3.79	10		
3.62	ms			3.63	2		
				3.48	2		
3.32	s	3.33	10	3.34	10	3.324	5
3.08	ms			3.07	3		
		3.00	3	3.02	3		
2.89	mw	2.90	3	2.91	2		
				2.82	1		
2.67	w						
2.57	vs	2.59	10	2.59	7	2.583	7B
2.47	w			2.47	2		
		2.42	5	2.44	4		
2.38	m			2.39	3	2.396	5B
2.25	mw			2.25	2		
				2.17	1		
2.14	m	2.15	2				
				2.12	1		
1.99	md	2.00	2	1.99	1	1.990	4B
		1.92	2	1.89	2		
1.71	vw					1.705	1
1.65	md			1.65	1	1.653	1
1.58	vvw						
1.50	s			1.50	1	1.508	4B

Impurities

D28, fraction 2–20 μm

d(A)	I	
6.64	1	ORTHOCLASE
6.53	2	O
5.88	2	O
4.24	8	O , Q
4.03	1	PLAGIOCLASE
3.952	3	O
3.871	1	O
3.788	7	O
3.674	1	P
3.627	1	O
3.560	1	O
3.465	5	O
3.325	10	O , Q
3.285	6	O
3.23	8B	O
3.193	2	P
2.996	6	O
2.931	1	O
2.907	3	O
2.767	2	

(fraction obtained by
sedimentation, dried
at 105°C)

(°) Brown l.c. pp.238-239 (+) < 2 μm, NH_4 form

(°°) oriented (°°°) disoriented

F6 : CuKα , 37 kV, 7 mA; D28 : CuKα , 34 kV, 19 mA

06 Chrysotile

lit.°		GB37	
d(A)	I	d(A)	I
7.36	10	7.32	100
4.58	6	4.50)B	13
		4.47)B	12
3.66	10	3.66	61
2.66	4		
2.594	4	2.59 B	6
2.549	6		
2.456	8	2.45 B	10
2.282	2		
2.215	2		
2.096	6	2.10 B	2
1.829	2	1.83 B	2
1.748	6	1.75 B	2
1.536	8	1.54	8
1.465	2	1.46	2
1.317	4		

(°) Brown l.c. p. 118

GB37: CuKα , 40 kV, 14 mA

  B: Broad

Impurity: swelling clay 19.0

07 Crocidolite

d(A)	I
9.21	1
8.36	100
4.88	
4.50	11
3.88	
3.421	
3.263	1
3.106	26
2.968	B
2.791	10
2.723	18
2.560	4
2.533	11
2.319	5
2.262	B
2.173	5
2.028	B
1.921	13
1.614	
1.602	11
1.509	
1.425	

08   Talc

lit.°		F3°°		F3°°°		F6°°		Impurities F3	
d(A)	I	d(A)	I	d(A)	I	d(A)	I		
9.30	70	9.44	80	9.42	100	9.6	10		
4.65	10	4.70	19	4.69	17	4.68	5		
4.57	80	4.54	2	4.57	1	4.57	6	14.02	1
3.10	70	3.125	100	3.123	90	3.14	10	7.06	2 chlorite
2.60	60	2.606	2	2.597	1	2.61	4	3.53	2
2.48	100	2.488	10	2.483	3	2.48	5	2.33	1
2.20	20	2.220	5			2.22	3		
2.10	10	2.106	5			2.11	3		
1.92	5								
1.86	20	1.872	8	1.872	6	1.87	4		
1.72	10								
1.67	5	1.676	7	1.674	4	1.68	3		
1.56	10	1.559	5	1.559	4	1.56	3		
1.52	70	1.527	4	1.526	1	1.52	3		
1.51	10	1.510	1						
1.46	10								
1.39	20	1.395	11	1.394	6	1.39	3		
1.33	10	1.336	2	1.336	3	1.33	3		

(°) X.P.D.F. card 3-0881
               3-0887
(°°) disoriented, ground to <20 μm
(°°°) oriented, ground to <15 μm

F3: CuKα , 30 kV, 20 mA
F3°°, R.H. 55%; F3°°°, R.H. 55%

10 Gibbsite

lit.° d(A)	I	JAP3 d(A)	I	d(A)	I	S3 d(A)	I	d(A)	I
4.85	100	4.85	100			4.853	100		
4.37	40	4.374	28			4.375	40		
4.31	20	4.326	17			4.322	20		
3.35	6	3.363	6			3.361	6		
3.31	10	3.329	10	2.702	0.5	3.316	10		
3.18	7	3.187	9	2.652	0.5	3.186	7		
3.10	4	3.110	3	2.467		3.109	3	2.468	9
2.451	15	2.455	20			2.453	12		
2.422	4	2.426	7			2.426	4		
2.382	25	2.388	19			2.386	19		
2.288	4	2.2902	4			2.2909	4		
2.244	6	2.2468	8	2.1910	1	2.2469	6		
2.165	8	2.1664	10			2.1662	7		
2.082	1	2.0858	1						
2.042	15	2.0499	16			2.0482	12		
2.024	1								
1.991	8	1.9968	12	1.9665	2	1.9943	8		
1.916	6	1.9187	10			1.9167	6		
1.801	10	1.8055	15			1.8042	9		
1.750	9	1.7526	15			1.7508	9		
1.685	7	1.6865	13	1.6458	0.5	1.6840	7		
1.655	2	1.6584	3	1.6383	1	1.6568	2	1.6369	1
1.590	2	1.5946	2	1.5876	2	1.5926	2	1.5861	2
1.574	1	1.5757	2						
1.555	1	1.5540	2						
1.533	1	1.5337	1						
1.485	1	1.4868	1						
1.477	1								
1.457	8	1.4576	9			1.4573	7		
1.440	4	1.4409	5			1.4404	4		
1.411	5	1.4116	6			1.4119		1.4070	4
1.402	4	1.40184	5			1.4019	3		
1.380	1	1.38102	1			1.38064	1		
1.361	2	1.36198	3			1.36165	2		
1.330	1	1.33131	2			1.33163	1		
1.319	1	1.31709	4			1.31711	1		
1.249	1	1.24943	3						
1.231	1	1.2329	1						
1.222	1								
1.213	2	1.21146	6			1.21552			
1.193	1	1.19261	3						
		1.18000	2						
		1.14515	2						
		1.12208	2						
		1.0940	0.5						

(°) Brown l.c. p.384

JAP3 : CuKα , 36 kV, 20 mA.
   ground to <37 μm, R.H. 64%

S3 : CuKα , 50 kV, 28 mA
   ground with 25 vol.% KCl

11 Magnesite

d(A) lit.°	I	d(A) B3	I
		3.54	2
2.742	100	2.738	100
2.503	17	2.501	20
2.318	4	2.320	3
2.102	43	2.102	24
1.939	12	1.938	6
1.769	3	1.770	3
1.700	34	1.699	30
1.510	4	1.509	1
1.488	5	1.486	2
1.426	4	1.405	2
1.371	3	1.370	1
1.354	7	1.353	5
1.338	8	1.337	5
1.252	3	1.250	6
1.2386	1	1.238	1
1.2022	1	1.202	1
1.1798	1	1.179	2
1.1583	1		
1.1297	1	1.100	1
1.1011	1		
1.0669	4	1.066	3
1.0510	1	1.050	1
1.0145	1	1.0137	1
0.9692	2	0.9685	1
0.9573	1		
0.9455	1	0.9450	1
0.9188	3	0.9183	1
0.9134	7	0.9133	6
0.8941	1	0.8938	1
0.8837	1	0.8839	1
0.8758	1	0.8751	1

(°) X.P.D.F. card 8-479

B3 : CuKα , 35 kV, 16 mA, Ni filter
ground to 5-15 μm, mixed with
dioctylphtalate and deposited
on a glass support; 40-45% R.H.

F13 : CuKα ,40 kV, 15 mA,
ground to <40 μm,

B3: Rhombohedral space group: R $\bar{3}$ c
Hexagonal indeces: a = 4.631 ± 0.002 A
c = 15.004 ± 0.005 A
Number of molecules in the unit cell Z = 6
Calculated density D_x = 3.014 g/cm^3

Impurities (F13):

d(A)	I	mineral
14.289		CHLORITE
12.762		
9.995		MICA
9.400	6	TALC
7.132	3	CHLORITE
4.984		MICA
4.746	3	CHLORITE
4.686		TALC
4.263		QUARTZ
3.554	3	CHLORITE
3.348	6	QUARTZ,MICA
3.118	5	TALC
3.037		CALCITE
2.889	20	DOLOMITE
2.829		CHLORITE
2.673		DOLOMITE
2.500		TALC(MAGNESITE)
2.458		QUARTZ
2.405		DOLOMITE
2.281		QUARTZ,CALCITE
2.214		TALC
2.195		DOLOMITE
2.069		DOLOMITE
2.018		DOLOMITE
1.998		MICA
1.983		QUARTZ
1.872		TALC,CALCITE
1.848		DOLOMITE
1.818		QUARTZ
1.805		DOLOMITE
1.788		DOLOMITE
1.674		TALC
1.635		MICA
1.566		DOLOMITE,TALC
1.542		DOLOMITE,QUARTZ
1.465		DOLOMITE

12 Calcite

lit⁰		B3	
d(A)	I	d(A)	I
3.86	12	3.86	6
3.035	100	3.038	100
2.845	3	2.835	2
2.495	14	2.498	7
2.285	18	2.287	12
2.095	18	2.096	11
1.927	5	1.928	3
1.913	17	1.910	18
1.875	17	1.877	14
1.626	4	1.624	1
1.604	8	1.601	5
1.587	2	1.588	1
1.525	5	1.524	4
1.518	4	1.518	3
1.510	3	1.510	2
1.473	2	1.472	1
		1.439	4
		1.420	3
		1.355	1
		1.337	1
		1.296	2
		1.284	1
		1.235	1
		1.179	1
		1.153	1
		1.142	1
		1.046	1
		1.044	2
		1.035	1
		1.011	2

Impurities:

F1 (of residue after treatment
with cold 0.1 N acetic acid)

9.9	5.02	1.98		mica muscovite
13.9	7.08	4.7	3.53	chlorite
4.26	3.34	1.81		quartz
2.70	2.42	2.20	1.91	1.61
				pyrite

(°) X.P.D.F. card 5-0586

B3 : CuKα , 40 kV,20 mA, Ni filter
    ground to 5-15 µm, mixed with
    dioctylphtalate, deposited in
    a layer on glass support.40-45% R.H.

B3: Rhombohedral, space group: R $\bar{3}$ c ; a = 6.374 A   α = 46°02'
    Hexagonal indeces: a = 4.985 A
                       c = 17.062 A
    Number of molecules in the unit cell Z = 6

13 Gypsum

lit.°		B3	
d(A)	I	d(A)	I
7.56	100	7.58	100
4.27	51	4.285	24
3.79	21	3.800	18
3.163	3	3.168	2
3.059	57	3.063	27
2.867	27	2.872	8
2.786	5	2.783	2
2.679	28	2.680	7
2.591	4	2.594	2
2.530	1	2.531	1
2.495	6	2.490	2
2.450	4	2.448	1
2.400	4	2.400	1
2.216	6	2.215	4
2.139	1	2.140	1
2.080	10	2.083	4
2.073	8	2.074	4
1.990	4	1.989	1
1.953	2	1.952	1
1.898	16	1.897	5
1.879	10	1.877	3
1.864	4	1.862	1
1.843	1		
1.812	10	1.811	3
1.796	4	1.795	1
1.778	10	1.777	3
1.621	6	1.619	3
1.599	1	1.599	1
1.584	2	1.582	1
1.532	1	1.578	1
1.522	1	1.520	1
1.50	)1		
1.48	)		
		1.457	1
		1.434	1
		1.364	2

Impurities:

B3 : After dissolution of
gypsum in a solution of
$Na_2S_2O_3$ one observes
calcite, dolomite, quartz
and traces of $SrSO_4$.

B3:Monoclinic system,   ASTM

a = 5.671 ± 0.006 A (5.68)

b = 15.186 ± 0.008 A (15.18)

c = 6.526 ± 0.006 A (6.51)

β = 118°29' ± 8'   (118°23')

Number of molecules in the
    unit cell: Z = 4

Calculated density: D_x = 2.315 g/cm^3

Observed density D_m = 2.32 g/cm^3

(°) X.P.D.F. card 6-0046

B3 : CoKα , 35 kV, 16 mA, Fe filter
    ground to 5-15 μm, 40-45 R.H.

OECD COMMENTS   K. Jasmund, Universität, Köln

The samples have been pretreated in two differerent ways. One
group of laboratories obtained the specimen <10 μm for investigation
by dry or wet grinding of the sample, whereas the other group split
the sample in different particle size fractions using the Atterberg
method. For clay minerals the latter method has the advantage of
enriching the main constituent in the finer particle size fractions.
With the first method, the presence of impurities (for instance feldspar,
particularly when the weak Co-radiation is used) is often missed, and
sometimes even d-values belonging to impurities are listed as belonging
to the main constituent. Also, grinding of the sample to a certain size
often seems to make it difficult to prepare an oriented specimen of
good quality.

Dispersion of the data

For ten of the minerals a sufficient number of data was reported to
allow a meaningful evaluation of the spread of the data.One can distinguish three groups according to the precision obtained:
For calcite, magnesite, crocidolite, and gibbsite:
d(mean)-d(measured) = $\pm$ 0.05 A to   <0.01 A
                 for   high   to medium/small d-values
For illite, attapulgite, kaolinite, and talc
d(mean)-d(measured) = $\pm$ 0.2   A to   ≤0.01 A
                 for   high   to   medium/small d-values
For montmorillonite and laponite
d(mean)-d(measured) = $\pm$ 3.0 A to $\pm$ 0.1 A to $\pm$ 0.01 A
                 for   high   to   medium to   small d-values.
For chrysotile and gypsum only two analyses each were reported.
The deviations for the two smectites, montmorillonite and laponite,
are very large. These cannot be attributed to measurement errors,
they must be a consequence of differences in the pretreatment procedures
and the conditions determining the state of hydration (or glycolation)
during the measurement.For these minerals it is very important to describe
in detail the methods of cation exchange and particle size fractionation,
the relative humidity and process of glycolation. Enclosure of the sample
during measurement is highly desirable.

Detailed comments on individual samples:

01 Montmorillonite.

Diffraction patterns were reported by 22 laboratories. The principal phase is a dioctahedral smectite, and according to the Hofmann-Klemen lithium test (F6, F3)   a montmorillonite.

The following impurities, with estimates of amounts present have been observed in the diffraction patterns:

Feldspar, specifically albite, 6-10%(3.21/3.22 A, 3.17/3.19 A)

Quartz, 3-5% (4.25 A, 3.34 A)

Mica, 2-3% (muscovite, biotite, 10.0 A, 5.0 A, 3.38 A)

Calcite, 1-2% (3.02/3.05 A)

Kaolinite, trace

The major part of the impurities may be removed by sedimentation.

02 Laponite

Patterns were submitted by 11 laboratories. The mineral is a synthetic tri-octahedral smectite of the hectorite type. Diffraction peaks are generally rather broad, indicating the small size of individual particles. No impurities are detected in the X-ray patterns.

03 Kaolinite (China Clay)

Twenty eight patterns were submitted. The kaolinite is well crystallized. The following impurities(with estimated amounts)were observed:

Mica, 8-12% (10.0;5.0;4.46;3.86;3.74;3.21;2.99;2.86;2.56;1.99;1.49 A)

Quartz,0.5-2% (3.34 A)

Feldspar, 2-3% (3.21 A)

Tourmaline, trace

Mixed layer mineral, trace

The impurities may be easily removed by sedimentation.

04 Attapulgite

Patterns were submitted by 12 laboratories. The principal phase is attapulgite, but there is a considerable amount of quartz in the sample, i.e. 22-26% according to quantitative analysis of the patterns by several laboratories. However, the quartz can be easily removed by sedimentation procedures. Impurities are :

Quartz, 22-26% (4.26;3.34;2.46;2.29;2.24;2.13;1.98;1.82;1.67;1.61; 1.54 A)

Dolomite (2.89 A) and Calcite (3.03; 1.91;1.87 A), 4%

Mica, (5.02;3.33;2.99 A), minor amount

Feldspar (3.99;3.19;3.10;2.68;2.51 A), minor amount
Chlorite (?)

05 Illite

The sample was examined by 15 laboratories. The principal phase
is Illite, $2M_1$.

Impurities are :

Quartz, 8% (4.26;3.34;2.47;2.12;1.99 A)

Feldspars (orthoclase and plagioclase), primarily in the large
particle size fraction (see table), in small amounts, together with
some quartz.

The impurities are easily removed by sedimentation, yielding an
almost pure illite.

06 Chrysotile

Two analyses were reported. The mineral is contaminated only
with a small amount of a swelling clay.

07 Crocidolite

According to seven analyses, the sample is a well crystallized crocidolite with a low content of impurities. (According
to IR analysis probably calcite). No table of data is presented.

08 Talc

Spectra were submitted by 8 laboratories. The following
impurities have been identified:

Chlorite, 10% (14.02;7.06;3.53;2.33 A)

Calcite (3.04;2.85;2.28 A)

Quartz (3.35 A)

The chlorite is difficult to remove from the sample.

10 Gibbsite

The gibbsite is very pure according to the 4 analyses submitted.
However, this synthetic material is less well crystallized than
some natural samples.

11 Magnesite

Spectra were submitted by 12 laboratories. The sample is very
impure and contains quartz, calcite, dolomite, mica, talc and
a chlorite. Spectral features attributed to these contaminants
are listed in the table.

12 Calcite

According to 7 analyses submitted, the sample is a well crystallized calcite. Impurities are small amounts of muscovite mica, chlorite, quartz, and pyrite. (see list of d-spacings attributed to these impurities in the table).

13 Gypsum

Impurities in this sample are calcite, dolomite, quartz, a trace of $SrSO_4$, and, according to IR a silicate.

CATION EXCHANGE CAPACITY

INTRODUCTION

The "cation exchange capacity" (C.E.C.) of a clay mineral
is generally understood to be equivalent with the layer charge.
This layer charge has a constant values as it is determined by
isomorphous substitutions within the crystallites. Hence, cation
exchange capacity is considered a material constant for a clay.

In order to determine this charge, the total amount of cations
in exchange positions on the clay surface must be measured. In most
methods the various cations which are present on the natural clay
surface are replaced by a single cation, and the total amount of
this cation species is then determined in the clay after washing
and isolating the clay. For example, the clay is treated with an
excess of ammonium acetate solution, and the ammonium content of
the washed clay is determined by means of the Kjeldahl method.
Since there is a danger that on prolonged washing ammonium ions are
displaced by hydrogen ions, an alcohol solution is used for washing.
Simultaneously, the alcohol keeps the clay in a flocculated conditions
which facilitates washing and filtration, or sedimentation.

In other methods the clays are converted to the barium form, and
the barium content of the washed clay is determined, for example
by conductometric titration with magnesium sulphate. Alternative
methods are based on the strong preference of the clay for the adsorption of organic cations, and the adsorbed amount of organic cation
is usually derived from the change in concentration of the organic
cation in the equilibrium solution.

When displacing the ions present on the clay surface with an
ammonium salt, the displaced cations may be determined in the
ammonium salt solution. Although information is thus obtained on the
exchange cations on the surface, one also determines cations of
soluble salts present in the clay, and the sum of the displaced
cations is then larger than the cation exchange capacity. On the
other hand, part of the cations in exchange position may be hydrogen ions. In this case the sum of the other cations which are usually
determined in the equilibrium solution will be less than the cation
exchange capacity.

The measured cation exchange capacity is only equivalent with
the layer charge if all charge compensating cations are accessible
for exchange. This is not true for illites in which the interlayer
cations are not exchangeable. The same is true for partially col-
lapsed smectites, e.g. potassium montmorillonite after heating or
after several drying and wetting cycles resulting in "potassium
fixation".

The constant charge concept has been accepted generally, but
was never proven for kaolinites. Alternative views on the origin
of the surface charge of kaolinites have been proposed and experimentally supported: according to these views the charge originates
in very thin surface coatings of alumina-silica (1).

Furthermore, edge surfaces of clay particles carry charges
due to the presence of broken bonds. The electrical double layers
on these surfaces are of the"constant potential"type, for which
the charge varies with the electrolyte concentration of the
equilibrium solution.The"constant potential" which is independent
of the electrolyte concentration, is a function of the pH of the
equilibrium solution, and may be either positive or negative.
Hence, the exchangeable ions associated with the edge surfaces
may be anions or cations, depending on pH. Therefore, if the
contribution of the edge surfaces can not be neglected with respect
to the contributions of the face surfaces, the total exchange
capacity will vary with electrolyte concentration, as well as with
pH. Then, the cation exchange capacity is no longer a material
constant. In the extreme, this is true for amorphous clays (allo-
phanes), which, therefore, have no fixed cation exchange capacity.

(1) Ferris,A.P. and Jepson,W.B. (1975) The exchange capacities
of kaolinite and the preparation of homoionic clays.
J.Coll.Interface Sci., 51, 245-259

CMS RESULTS

Cation exchange capacities were determined in three laboratories,
each using a different method. In order to obtain comparable results,
the particlants were asked to determine the C.E.C. on the samples
as received, and to determine the reference dry weight of each clay
by measuring the weight loss upon drying at 140°C for 2 hours on a
different portion of the sample. The data presented refer to this
dry weight.

The participants, and the methods used were the following:

A.E.Worthington, Chevron Oil Field Research Company, La Habra, Cal.
Method: Exchange with $BaCl_2$ and washing in a dialysis bag.
Determination of exchangeablebarium by conductometric titration with
magnesium sulphate.Determinations made in duplicate.
References: Worthington,A.E., Geophysics 38, 140-153, 1973.
Mortland,M.M. and Mellor,J.L., Proc.Soil Sci.Soc.Amer.,18, 363-364,1954.

W.T.Granquist, Baroid Division, NL Industries, Houston,Texas.
Method: Ammonium acetate exchange, Kjeldahl determination of NH_4
on the washed clay. Single determinations.

C.V.Clemency, State University of New York at Buffalo, N.Y.
Method: Ammonium acetate exchange, determination of the ammonium
content of the washed clay with an ammonia electrode.Duplicate runs.
Reference:Busenberg,E. and Clemency,C.V.,Clays and Clay Minerals,
21, 213-217, 1973.

CMS COMMENTS

There is reasonable agreement between the values obtained by
the three different laboratories using different methods, i.e.
results are generally within 5% of the mean.An exception are
the results for the synthetic mica-montmorillonite. This clay
contains primarily ammonium ions in exchange positions. Those
ammonium ions which are compensating the charges in the mica
layers of the interstratified mica-montmorillonite are probably
not accessible for exchange with barium cations, and this would
explain the discrepancy between the results obtained with the
ammonium methods and with the barium method.

CMS CATION EXCHANGE CAPACITY

code	name	Worthington		Granquist		Clemency		mean CEC	Ca	Mg	K	Na	Σ
		CEC	loss	CEC	loss	CEC	loss						
KGa-1	Kaolinite,well-crystallized	1.9	(0.45)	2.4	(0.39)	1.70		2.0					
KGa-2	Kaolinite,poorly crystallized	3.3	(1.21)	3.8	(0.90)	2.80		3.3					
SWy-1	Montmorillonite Wyoming	79	(3.62)	76.3	(7.43)	74		76.4	53.7	21.7	2.2	45.5	123
STx-1	Montmorillonite Texas	80	(4.00)	85.2	(14.0)	88		84.4	54.5	18.0	0.5	9.9	82.9
SAz-1	Montmorillonite Arizona	112	(5.89)			129	(9.0)	120					
SHCa-1	Hectorite California	46	(3.34)	45.6	(4.00)	40	(4.0)	43.9	95.8	11.3	1.3	37.5	146
Syn-1	Synthetic mica-montmorillonite	67	(1.92)	161	(4.96)	116							
PFl-1	Attapulgite, Florida	16	(6.96)	25.4	(10.7)	17		19.5	14.0	58.0	0.78	0.22	73

CEC expressed in meq per 100 g of dry clay; "loss": percent weight loss on drying 2 hours at 140°C.

Clemency determined the following values for the NBS clays:

NBS 97a : 6.8 meq/100g   NBS 98a : 16 meq/100g

According to the analysis for the cations in the ammonium acetate
solution after exchange, the Wyoming montmorillonite, the hectorite,
and the attapulgite seem to contain soluble salts, since the sum of the
exchange – cations exceeds the cation exchange capacity.

OECD RESULTS

Most of the laboratories used the ammonium acetate method. D5 and D24
applied the Mehlich method which is based on exchange with barium.
(see: Mehlich, A., Soil Sci., 66, 429-445, 1948).

One laboratory (P3) used a method developed by Davidson and Scheeler
(see: Davidson, D.T. and Scheeler, J.B., in " Symposium on Exchange
Phenomena in Soils, ASTM, New York, 1952).

OECD COMMENTS

The spread in the data is relatively large, even when discarding
some data (shown in brackets in the table), which fall far outside the
range of most of the reported data. This spread is probably primarily
due to variations in the reference dry weights used by the different
laboratories, and possibly to other factors.

The sum of the exchange-cations in the ammonium acetate solutions
is in most cases close to the cation exchange capacity, except for
attapulgite, which appears to contain soluble calcium.

OECD CATION EXCHANGE CAPACITY

No.	Name	B1	F6	D5°	D7	D24°	D28	P3°°	GB5	mean
01	Montmorillonite	80.25	87		95		75 74	84.0 83.2	(113)	82.5
02	Laponite			75.1		71.4				73.3
03	Kaolinite (China Clay)	(2.56)		5.3			1.45 1.50	(2.0) (1.9)	4.5	4.9
04	Attapulgite				8.3				9.7	9.0
05	Illite	26.8			23.0 22.8	29.2	(37.4) (38.4)		31.4	26.6

(°) Mehlich method

(°°) Method reference: Davidson,D.T. and Scheeler,J.B., Symposium on exchange Phenomena in Soils,
A.S.T.M., New York, 1952

Others used ammonium acetate methods.

Values in brackets were discarded in calculation of mean value.

OECD CATION EXCHANGE CAPACITY

No.	Name		H	Ca	Mg	K	Na	Σ	CEC
01	Montmoril-lonite	D5	0	28.0	34.0	2.0	32.8	96.8	82.5
		D7	–	62.5	14.0	2	33	111.5	
		GB5	–	44.7	6.7	2.5	35.0	88.9	
02	Laponite	D5	0	7.0	0.2	0.4	85.5	94.1	73.3
03	Kaolinite (China Clay)	D5	3.0	1.2	0.8	0.2	0.2	5.4	4.9
		GB5	–	1.02	0.72	0.12	0.28	2.14	
04	Attapulgite	D7	–	49	4.5	0.1	0.6	54.2	9.0
		GB5	–	30.3	6.8	0.6	1.5	39.2	
05	Illite	D7		18.75	11.88	2.15	0.25	33.03	26.6
		GB25		17.4	1.88	1.82	1.06	22.16	

SURFACE AREA

INTRODUCTION

Choice of methods, references

The most commonly used procedure for determining surface area
of a powder is to derive the amount of adsorbed nitrogen (or other
inert gas) at monolayer coverage from a BET plot of adsorption isotherm
data. Knowing the projected cross sectional area per molecule in a
monolayer, the surface area is calculated from the monolayer coverage.

Since inert gases in general do not penetrate between layers of
an expanding clay, only the total external surface area is determined
for these minerals. Methods based on the adsorption of polar molecules, such as water, glycol or glycerol, which are able to penetrate between the layers of expanding clays yield data for the total
internal and external surface areas. Combining both methods, external
and internal surface areas can be evaluated separately.

Although this approach is simple in principle, there are many
pitfalls when applying the methods to real systems. On one hand,
expanding clays do under certain circumstances admit nitrogen between
the layers, for example when the layers are separated by water molecules at less than monolayer coverage. On the other hand, for ionogenic
surfaces, such as clay surfaces, monolayer coverage is to a certain
extent dependent on number and kind of ions on the surface, hence in
reality one can not assume a fixed cross sectional area per molecule
at monolayer coverage which can be universally applied in calculating
surface area. Finally, the point at which the surface is just covered
by a monolayer is not always easy to establish unequivocally. A survey
of methods and a discussion of their limitations is given in reference 1.

For large grain size powders in the collections, surface areas were derived from gas permeability measurements (reference 5)

Calculation of (external) surface areas from adsorption isotherms

The specific surface area a_s is calculated from the monolayer capacity V_m , the amount of gas expressed as a volume at s.t.p. (0° C and 1 at. or 273.15 K and 101.325 kPa) needed to cover the surface with a single monolayer, using the equation:

$$a_s = \frac{V_m a(N_2) N}{V_o W}$$

where N is Avogadro's constant ($6.0247 \times 10^{23} mol^{-1}$)

a(N_2) is the cross-sectional area of nitrogen (chosen as $16,2$ A^2 if the layer is assumed to be close packed)

V_o is the volume of one mole of ideal gas at s.t.p.

($22\ 410\ cm^3 mol^{-1}$)

W is the weight of the sample

Inserting numerical values gives:

$$a_s/m^2 g^{-1} = 4.355 \times \frac{V_m/cm^3}{W/g}$$

If Argon is used as the adsorbate, using a(Ar) = 18.2 A^2 the numerical factor in the above equation is 4.982. For water with a(H_2O) = 11.0 A^2 the factor is 2.957.

The monolayer capacity V_m is determined from the BET equation:

$$\frac{p}{V(p^o - p)} = \frac{1}{V_m c} + \frac{c - 1}{V_m c} \frac{p}{p^o}$$

where p is the equilibrium pressure of the adsorptive,

p^o is the saturation pressure of the adsorptive at the temperature of adsorption,

c is a constant,

V is the volume of gas adsorbed, calculated at s.t.p.

If, when $p/V(p^o - p)$ is plotted against the relative pressure p/p^o; a straight line is obtained, V_m and c may be calculated from the gradient and intercept. The range of p/p^o in which a linear region is observed experimentally is often severely restricted, and is often much shorter than the relative pressure range of 0.03 to 0.30 conventionally used.

In rare cases a Langmuir isotherm is obtained, requiring a
different way of plotting the data.

An interlaboratory study of the BET nitrogen adsorption method
on a suite of potential surface area standards has been conducted
under the auspices of the Society of Chemical Industry in the U.K.,
the International Union of Pure and Applied Chemistry, and the
National Physical Laboratory in the U.K. (reference 2). Four of
the eight samples studied were found to be suitable surface area
standards. Samples of these may be ordered from the National
Physical Laboratory, Teddington, Middlesex, U.K. The analysis of
the results of the study and advice on the proper conduct of the
experimental determination and analysis of the adsorption isotherms are given in reference 2. The principal requirement for
obtaining reproducible results isthat the pretreatment and outgassing procedures must be reproducibly controlled. This is most
important when the state of hydration of the surface is very
sensitive to prolonged exposure to humidity, as in swelling clays
and particularly for attapulgite and sepiolite for which water
content, as well as crystal structure, and therefore the measured
areas are strongly dependent on outgassing temperature (reference 4).

Data are presented for nitrogen or argon, water and glycerol
adsorption methods. The first three are based on BET analysis of
adsorption isotherms. For the glycerol adsorption method a thermo-
gravimetric method was used in which the monolayer coverage is
established from the occurrence of a plateau of nearly constant
weight in a narrow temperature range in which the monolayer complex is apparently relatively stable.(reference 3).

From glycerol adsorption data, total surface area is calculated as
follows: The glycerol monolayer is assumed to have a thickness
of 4.5 A as shown by the increase of the c-spacing of swelling
clays at monolayer coverage, and a specific volume of 0.885 cm^3/g
in the temperature range in which the monolayer complex is stable, i.e.
about 200 °C. Hence, 1 gram of glycerol at monolayer coverage will
cover an area of 1967 m^2, and a retention of 1 g of glycerol per
100 g of clay is equivalent to a surface area of 19.67 m^2/g. For
n % retention on a non-swelling clay the area is n x 19.67 m^2/g,
and for an expanding clay the sum of the external and internal
surface areas is equal to: (n x 2 x 19.67 - BET nitrogen area).

References

(1) van Olphen,H., Determination of surface areas of clays –
    Evaluation of methods. IUPAC, Supplement to Pure and Applied
Chemistry: "Surface Area Determination", Butterworths, London,
1970, pp. 255-271

(2) Everett,Douglas H., Parfitt,Geoffry D., Sing, Kenneth S.W.,
    and Wilson,Raymond, The SCI/IUPAC/NPL Project on Surface

Area Standards, J.appl.Chem.Biotechnol., 1974, 24, 199-219

(3) Madsen,F.T., Surface Area Determination of Clay Minerals by
    Glycerol Adsorption on a Thermobalance. Thermochimica Acta,

(4) Alvarez, Fernandez, T. Superficie Especifica y Estructura
    de Poro de la Sepiolita Calentada a Diferentes Temperaturas,
Anales Reunion Hispano-Belga de Minerales de la Arcilla, CSIC,
Madrid, 1970, pp.202-209.

(5) Permeability method

This determination of the specific surface area of powders
is carried out with the Griffin apparatus, and is based on Rigden's
permeability method, relating the velocity of air through a compressed powder at a given pressure gradient, the density of the
solid, and the void volume,with the total surface area of the
particles.

The stainless steel permeability cell has a cross section A
$(=5.0645\ cm^2)$. The powder is uniformly compressed by manually
operating a piston,to a thickness L, measured to $\pm$ 0.01 mm with
a vernier comparator. The weight of the sample is W, and its grain
density ρ (determined pycnometrically).

In the apparatus one measures the velocity of air through the
sample from the flow velocity of a liquid in a calibrated U-tube.

The specific surface area S is calculated from the expression:

$$S^2 = \frac{2\,A\,\varepsilon^3\,g\,d}{k(1-\varepsilon)^2\,\eta\,\rho^2\,L\,a} \cdot \frac{T \cdot h_1^{h_2}}{\log_e h \cdot \frac{h_1}{h_2}}$$

in which ε = the ratio of void volume to total volume $(=1-W/\rho AL)\rho$

$\qquad$ g = the gravitational constant

$\qquad$ d = the density of the liquid in the calibrated tube

T = the duration of the liquid flow in the tube between h_1 (the start of the experiment) and h_2

k = the Kozeny constant (= 5.0)

η = the viscosity of the air

a = the cross section of each branch of the U-tube

The results are accurate to ± 5%

References:

Kozeny, Ber.Wien.Akad., 1927, 136 a, 271

Carman, Trans.Inst.Chem.Eng., 1937, 15, 150

　　　　　J.S.C.I., 1938, 57, 225

Lea and Nurse, J.S.C.I., 1939, 58, 277

Rigden, J.S.C.I., 1943, 62, 1

Rigden, J.S.C.I., 1947, 66, 130

CMS RESULTS

(a)　BET Method, N_2 -

　　　　R.Wilson, Division of Inorganic and Metallic Structure,

　　　　　　National Physical Laboratory, Teddington, Middlesex, U.K.

Relative pressure range: 0.05-0.25 , six points.

Indicated errors are standard deviations from least squares analysis.

One run on one sample of each clay.

Pretreatment, see table.

code	name	outgassing °C/ h / /torr	weight loss %	area $m^2 g^{-1}$	BET c
KGa-1	Kaolinite, well-crystallized	193/16 / 4×10^{-5}	0.68	10.05 ± 0.02	131
KGa-2	Kaolinite, poorly crystallized	199/16 / 4×10^{-5}	1.3	23.50 ± 0.06	127
SWy-1	Montmorillonite, Wyoming	196/15.5 / 5×10^{-5}	5.3	31.82 ± 0.22	838
STx-1	Montmorillonite, Texas	198/15 / 2×10^{-5}	14.4	83.79 ± 0.22	138
SAz-1	Montmorillonite, Arizona(Cheto)	198/15 / 2.5×10^{-5}	14.1	97.42 ± 0.58	537
Syn-1	Synthetic mica-montmorillonite	196/15 / 2×10^{-5}	5.5	133.66 ± 0.72	93
SHCa-1	Hectorite, California	200/15 / 2.5×10^{-5}	2.7	63.19 ± 0.50	782
PFl-1	Attapulgite, Florida	199/15 / 2.5×10^{-5}	15.7	136.35 ± 0.31	123

(b)   BET Method, N_2

       J.Thomas,Jr., R.R.Frost, B.F.Bohor, Illinois State Geological
       Survey, Urbana, Ill., USA.

Relative pressure range: 0.15-0.25 , three points.

Straight lines drawn visually.

One run on 3-5 different samples of each clay.

Pretreatment: Degassed overnight at 200°C at less than 10^{-2} torr.

code	name	individual runs m^2/g					mean m^2/g
KGa-1	Kaolinite,well-crystallized	8.69	8.77	8.79			8.75
KGa-2	Kaolinite,poorly crystallized	21.7	19.9	20.2			20.6
SWy-1	Montmorillonite, Wyoming	32.2	30.6	30.2			31.0
STx-1	Montmorillonite, Texas	79.0	82.2	80.6	78.9		80.2
SAz-1	Montmorillonite, Arizona(Cheto)	94.1	93.4	94.4	93.6		93.9
Syn-1	Synthetic mica-montmorillonite	135.3	139.9	129.2	138.7	135.0	135.6
SHCa-1	Hectorite, California	64.7	64.0	66.6	62.0		64.3
PFl-1	Attapulgite, Florida	123.8	131.0	126.5			127.1

(c) BET Method, N_2, and Glycerol Method

F.T.Madsen, Laboratory for Clay Mineralogy, Institute for
Foundation Engineering and Soil Mechanics, Swiss Federal
Institute of Technology, Zürich, Switzerland.

Pretreatment: Amorphous coatings of iron oxide were removed by the
method of Mehra and Jackson (Clays and Clay Minerals, _7_, 317-327,
1960). The clay was subsequently saturated with Ca and freeze-dried.
For the BET determination the clays were degassed overnight at
150°C at 10^{-6} torr.

code	name	percent retention glycerol	m^2/g area	m^2/g area N_2
KGa-1	Kaolinite, well-crystallized	0.81 ± 0.11	16 ± 2	10
KGa-2	Kaolinite, poorly crystallized	1.26 ± 0.21	25 ± 4	24
SWy-1	Montmorillonite, Wyoming	17.61 ± 0.57	662 ± 22	31
STx-1	Montmorillonite, Texas	17.30 ± 0.63	599 ± 24	82
SAz-1	Montmorillonite, Arizona(Cheto)	23.19 ± 0.44	820 ± 18	92
SHCa-1	Hectorite, California	13.83 ± 0.30	486 ± 12	58

(d)   BET One-Point Method, N_2
       C.V.Clemency and E.Busenberg, State University of New York,
       Buffalo, N.Y., USA

To evaluate the applicability of simple and quick routine methods
based on the one-point BET method, some measurements were taken with
a commercial instrument: Monosorb Model MS-3, Quantachrome Corporation,
Greenvale, N.Y. Readings were taken at 0.3 relative pressure in
adsorption-desorption cycles. Averages of three cycles on one sample
of each clay and reference material to compute surface area.
Pretreatment: There is no provision for degassing the samples under
vacuum and heating. However, the samples were heated overnight at
140°C at atmospheric pressure prior to the measurement.

Gasil Silica I (NPL)	286.1 ± 3.5 (NPL)	258
TK 800 Silica (NPL)	165.8 ± 2.1 (NPL)	163
Duke Silica 209 (°)	24.3 (Duke)	25.5
NBS 97a, Flint Clay	---	15.0
NBS 98a, Plastic Clay	---	30.3
KGa-1 Kaolinite,well-crystallized	10.05 (NPL)	9.21
KGa-2 Kaolinite,poorly crystallized	23.50 (NPL)	21.9
SWy-1 Montmorillonite, Wyoming	31.82 (NPL)	134.3
STx-1 Montmorillonite, Texas	83.79 (NPL)	69.3
Syn-1 Synthetic mica-montmorillonite	133.66 (NPL)	103.8
PFl-1 Attapulgite, Florida	136.35 (NPL)	212.2

(°) Duke Surface Area Standard 209, Duke Standards Co.
    Palo Alto, California, USA

CMS COMMENTS

The data obtained by NPL are considered best values, however those reported by Thomas et al and by Madsen based on the BET nitrogen method are close to those of NPL. The one-point method appears to give good values only for the kaolinites (as well as for the two lower area silicas); the data for the swelling clays and attapulgite are erratic. This is probably due to the presence of residual water in the samples, and perhaps also a consequence of the usually considerable hysteresis in the adsorption-desorption cycles for these clays.

All reported data, except for the one-point data for the swelling clays, do not deviate more than 10% from the mean for each sample, and in many cases they are within 5% of the mean.

The total (glycerol) surface area data are of the expected order of magnitude (about 750 m^2/g for smectites) taking into account the presence of impurities which reduce the crystallographic area calculated for the pure clay, particularly in the case of hectorite which contains a considerable amount of calcite. The value for the Arizona montmorillonite seems too high.

OECD RESULTS

Although nine laboratories participated in the determination of surface areas, no systematic effort was made and only some 30 data points were reported, using a variety of methods and equipment.

Primarily, the BET method was applied, in one laboratory with argon rather than nitrogen. One laboratory used a routine method based on desorption, and one laboratory applied the permeability method to coarse grained samples.

Most data were obtained on the samples as received, but some data were obtained on size-fractionated samples. Drying and degassing procedures varied widely. In most cases only one run was made on a given sample.

A few data on total area were taken using water as the adsorptive.

The reported data and a brief description of the conditions of measurements and equipment used are collected in the table.

No.Name	lab. area m²/g	pretreatment	method	equipment	mean m²/g
01 Montmoril-lonite	F2	44.5	degassed 18 h at 60°C and at 10^{-4} torr	BET, N_2		46.0
01	F3	48.15	degassed overnight at 105°C and at 10^{-3} torr	BET, N_2 Six points p/p_0:0.034-0.185	Own construction	
01	D22	48.87	not specified	BET, N_2	Areameter, Firme Ströhnlein,Düsseldorf	
01	GR3	42.5 ±1.0	dried at 100°C,degassed at 95°C at 8×10^{-3} torr	BET, Ar, data on three samples	Atlas BETograph, Firme MAT, Bremen	
01	GB5	35.8	dried overnight at 160°C, degassed 4h in N_2 atm.	Desorption of N_2	Sorptometer Perkin-Elmer	
01		20.6	dried one week at 160°C			
01	F3	86.14	converted to H-clay, size fraction <2 μm,degassed overnight at 105°C and at 10^{-3} torr	BET, N_2 Four points p/p_0:0.091-0.121	Own construction	
02 Laponite	F3	360	degassed overnight at 105°C and at 10^{-3} torr	BET, N_2 Six points p/p_0:0.086-0.207	Own construction	
03 Kaolinite (China Clay)	F3	7.17	degassed overnight at 105°C and at 10^{-3} torr	BET, N_2 Four points p/p_0: 0.126-0.204	Own construction	6.6
03	D7	6.6	not specified	BET, N_2	Areameter	
03	D22	6.91	not specified	BET, N_2	Areameter, Firme Ströhnlein, Düsseldorf	
03	S5	5.79	dried 12 h in air at 80°C	BET, N_2 duplicate runs one sample: 5.53 6.06	not specified	
03	GB5	5.6	dried overnight at 160°C degassed 4h in N_2 atm.	Desorption of N_2	Sorptometer Perkin Elmer	
03	B1	8.81	size fraction < 10 μm dried overnight at 105°C degassed overnight at 110°C below 10^{-6} torr	BET, N_2	Own construction	
03	F3	13.22	converted to H-clay,size fraction <2 μm,degassed overnight at 105°C and at 10^{-3} torr	BET, N_2 Five points p/p_0: 0.143-0.261	Own construction	

No.Name	lab.	area m²/g	pretreatment	method	equipment	mean m²/g
04 Attapulgite	F2	83.1	degassed 18h at 60°C and at 10⁻⁴ torr	BET, N_2	not specified	
04	D7	53	not specified	BET, N_2	Areameter	
04	D22	49.7	not specified	BET, N_2	Areameter, Firme Ströhnlein,Düsseldorf	
04	GB5	72.9	dried overnight at 160°C degassed 4h in N_2 atm.	Desorption of N_2	Sorptometer, Perkin-Elmer	
04	B1	65.5	size fraction <10 µm dried overnight at 105°C degassed overnight at 110°C below 10⁻⁶ torr	BET, N_2	Own construction	
05 Illite	D7	89.7	not specified	BET, N_2, 3 sample runs: 89.9 89.0 90.3	Areameter	101
05	D22	111.9	not specified	BET, N_2	Areameter, Firme Ströhnlein,Düsseldorf	
05	GB5	89.0	dried overnight at 160°C degassed 4h in N_2 atm.	Desorption of N_2	Sorptometer, Perkin-Elmer	
08 Talc	F2	2.37	degassed 18 h at 60°C and at 10⁻⁴ torr	BET, N_2	not specified	2.37
08	B1	5.75	size fraction <10 µm dried overnight at 105°C degassed overnight at 110°C below 10⁻⁶ torr	BET, N_2	Own construction	

No.Name	lab.	area m²/g	pretreatment	method	equipment	mean m²/g
10 Gibbsite	B1	8.16	dried overnight at 105°C degassed overnight at 110°C below 10^{-6} torr	BET, N_2	Own construction	7.3
10	GR3	6.56	dried at 100°C, degassed at 95°C at 8×10^{-3} torr	BET, Ar	Atlas BETograph, Firme MAT, Bremen	
11 Magnesite	B3	0.094		Permeability method	Griffin apparatus	0.094
11	GR3	0.42	dried at 100°C, degassed at 95°C at 8×10^{-3} torr	BET, Ar	Atlas BETograph, Firme MAT, Bremen	
12 Calcite	B1	<0.5	dried overnight at 105°C degassed overnight at 110°C below 10^{-6} torr	BET, N_2	not specified	0.067
12	B3	0.067		Permeability method	Griffin apparatus	
13 Gypsum	B3	0.075		Permeability method	Griffin apparatus	0.075

Surface areas from water vapor adsorption isotherms

No.Name	lab.	area m²/g	pretreatment	method	equipment	mean m²/g
02 Laponite	B1	299.6	degassed overnight at 110°C below 10^{-6} torr	BET, H_2O p/p_0:0.05-0.35	Own construction	
03 Kaolinite (China Clay)	B1	13.5	size fraction <10 μm as above	as above		
04 Attapulgite	B1	172	as above	as above		
08 Talc	B1	6.94	as above	as above		

OECD COMMENTS

Rapporteur: Th.Skulikidis, Physical Chemistry and Applied
            Electrochemistry, National Technical University,
            Athens, Greece.

Underligned values in the table were used to calculate a mean
value which is reported in the summary, Part I. In general, data
are so sparse and taken under such a variety of conditions that no
firm conclusions can be drawn on effects of pretreatment, or methods
and equipment used.

The following are some detailed comments on individual sample data.

01 Montmorillonite

Values from four laboratories are reasonably close and yield a
mean of 46.0 m^2/g with deviations of up to 8% of the mean. The
desorption method gives low values. Prolonged drying at 160°C
reduces the area accessible to nitrogen considerably.

02 Laponite

The nitrogen area of this synthetic clay is very large, about
half the theoretical crystallographical area of smectites, indicating a very small degree of layer association. The area determined with water as the adsorptive is also large,though somewhat
smaller than the nitrogen area for reasons that are not understood.

03 Kaolinite (China Clay)

Results from four laboratories are reasonably close and yield a mean
of 6.62 m^2/g with deviations of up to 12% from the mean value.
The desorption method again yields a low value. Higher values are
obtained for smaller particle size fractions. The area of one of
these fractions is 8.81 m^2/g for nitrogen adsorption, and 13.5
m^2/g for water adsorption,although these should be expected to be
equal for a non-swelling clay.

04 Attapulgite

Nitrogen areas vary between about 50 and 83 m^2/g. This large spread
in the data is undoubtedly due to variations in degassing conditions for
which the measured area is very sensitive because of structural changes
occurring in the mineral with a change in temperature, and because of
changes in the state of hydration of the sample.

An area of 172 m^2/g was reported using water as the adsorptive.

05 Illite

Data from three laboratories are reasonably close and yield a mean value of 96.9 m^2/g with deviations of up to 15.5 % from the mean value. Eliminating the results from desorption experiments, which are probably low, the mean value for the other two determinations is 100.8 m^2/g with a deviation of 11 % from the mean value.

ELECTRON MICROSCOPY AND DIFFRACTION

A. Oberlin

INTRODUCTION

The electron microscope may be used in two different ways: either
as a microscope proper, or as a high resolution diffraction apparatus.
(1,2)

In the latter application the apparatus serves to generate a parallel
monochromatic electron beam. The diffraction technique, like X-ray
diffraction is a statistical method. The electron diffraction technique
has certain advantages, e.g. it requires only small amounts of material.
Also, it is often possible to obtain better orientation in the sample
since only small quantities of material are needed. The diffraction
technique provides structural information on the total sample, and
is helpful in identification of the mineral.

Application of the apparatus as a microscope comprises dark field
and bright field microscopy as well as "selected area diffraction"
(SAD). The first two provide morphological information since single
particles can be discerned separately with proper sample preparation
techniques. SAD provides information on crystal imperfections in
single particles.

Particularly in those cases where X-ray diffraction gives insufficient information because of the small size of the particles and
imperfect crystal structures,yielding band diagrams, the combination
of SAD and high resolution electron diffraction provides a more
precise analysis of the crystalline structure of the material(3,6)

Morphological information is important for the understanding of
the genesis and diagenesis of clay minerals (4) and of the rheo-
logical properties of clay-liquid systems. (5,10c)

217

Recommendations for sample preparation(7,10c)

Only the simpler methods will be considered, and for example the
use of ultramicrotomes will not be discussed.

Solid specimens of phyllites

Investigation of particle association in the crude mineral should
be done by means of replicas, preferably using the technique developed
by Bates and Comer (8).

Dispersed particle specimens

a.Preparation of suspensions.

Generally speaking, phyllites with a sufficiently high cation
exchange capacity (> 15 meq/100g) can be satisfactorily dispersed in
water if they are prepared in the Na-form. Natural specimens which
are likely to contain various impurities and which are often in the
Ca-form should be subjected to two 2-hour treatments with 1 N NaCl, after
which they should be repeatedly washed and centrifuged until the clay
remains in suspension.When the sample contains iron oxides, it is
recommended to use the treatment developed by Mitchell and Mackenzie (9).

Microcrystalline material with zero or low exchange capacity
usually are satisfatorily dispersed by raising the pH of the suspension
to about pH 9 with ammonia.

b. Deposition of specimens on holder grids.

Preferably the thinnest possible carbon film support should be used
to deposit the specimen. After suitably diluting the suspension and
placing a drop on the specimen holder, the mesh should be dried in
a dessicator over silica gel. The proper concentration of the suspension
must be determined by trial and error, since that concentration is
dependent on such factors as particle dimensions and particle charge.
Suitable concentrations vary from about 0.5 percent, for example for
kaolinites, to as low as 1 ppm for smectites. Preparations of very
thin particles should be shadow cast with chromium or carbon-platinum
alloy. In the case of smectites (montmorillonite, hectorite etc.)
the method of specimen preparation has a very large bearing on the
observed morphology of the mineral (See references 4,5,10a,10b,10c).

c. Oriented deposits (for high resolution electron diffraction)

Deposits are called perfectly oriented when the planes (001)
of the lamellar particles are all perpendicular to the same axis.
In order to obtain a diagram of sufficiently high intensity, the
deposit must be composed of several superimposed layers of
particles. Rather than placing a single concentrated drop on the

mesh, it is preferable to use several drops of a very dilute
suspension, allowing each one to dry before applying the next.

The following notes apply to the various individual specimens:
Montmorillonite and Laponite

For these specimens replicas are not very useful.

The main problem is to obtain a deposit on the speciman holder
grid that gives a faithful representation of the individual particles;
Hence, it is necessary to start with a very well-dispersed suspension,
and electrical deposition techniques are recommended. It is often
helpful to charge the supporting film by adsorption of an organic
cation. The resulting positive charge of the membrane improves the
deposition of the negative plates in a flat position on the membrane,
and often prevents the curling of the plates.

SAD, and perpendicular and oblique high resolution electron
diffraction are useful for studying the atomic structure of the
primary particles. The type of disorder in random layer stacking may
be studied by combining SAD and dark field micrographs from the
different diffracted beams.

Kaolinite (China Clay), and Illite

Replicas of the crude specimens are useful for studying particle
association in the mineral.

The study of the morphology of the particles deposited from
dilute suspensions is of particular interest in relation with
the degree of crystalline order in the kaolinites, and in
relation with genesis and diagenesis of the illites.

Attapulgite

Both replica studies and observation of the morphology of particles
deposited from suspensions are of interest, however, it is often
difficult to obtain a stable suspension of this material.

Precautions

Since electron microscopy is not a statistical technique, but is
adapted to the study of individual particles, it is highly sensitive
to the presence of impurities. This is particularly important in
work on phyllites since these are often present as contaminant in
the air, therefore, the entire area of at least two specimen holder
grids should be examined.

Certain problems and limitations of the techniques are inherent
in the equipment. Since the specimen holder is in vacuum it is
impossible to work on hydrated minerals , for example halloysite,
or on thosewhich are liable to decomposition.

The electron beam may cause contamination of the specimen,
therefore one should use either a specimen holder cooled with
liquid air, or an anti-contaminant trap, or a specimen holder
heated to about 100°C, or one provided with adjustable leak.

The beam may cause local hot spots in specimens that are poor
heat conductors, or the beam may rupture bonds due to ionisation,
etc. Phyllites in particular are easily damaged in the beam, and
the resulting corrosion patterns which are readily visible may lead
to incorrect interpretations of morphology.

Referring to the problems in preparing specimens of smectites,
special care in sample preparation should be taken and preferably
results from several different preparation techniques should be
compared.

REFERENCES

1.   Oberlin, A., Electron microscopy and diffraction
                applied to the study of random layer lattices.
                RCA Scient. Instrum. News, 12, 2, 6, 9, and
                12, 3, 8-13, 1967

2.   Tchoubar, C. et Oberlin, A., Application de la
                microscopie et de la diffraction électroniques
                à l'étude des minéraux microcristallisés.
                J. Microscopie, 6, 541-556, 1967

3.   (a) Oberlin, A. et Mering, J., Observations en
                     microscopie et microdiffraction électroniques
                     sur la montmorillonite Na.
                     J. Microscopie, 1, 107-120, 1962

                (b) Oberlin, A. et Mering, J., Observations sur
                     l'hectorite. Bull. Soc. Franc. Miner. Crist.,
                     89, 29-40, 1966

                (c) Mering, J. and Oberlin, A., Electron optical
                     studies of smectites.
                     Clays and Clay Minerals, 15, 3-25, 1967

4.   (a) Oberlin, A., Freulon, J.M. et Lefranc, J.Ph.,
          Etude minéralogique de quelques argiles des
          grès de Nubie et du Fezzan (Lybie). Bull.
          Soc. Franc. Minér. Crist., 81, 1-4, 1958

      (b) Oberlin, A. et Freulon, J.M., Etude minéralogique
          de quelques argiles des séries primaires du
          Tassili m'Ajjer et du Fezzan. Bull. Soc. Franc.
          Minér. Crist., 81, 186-189, 1958

      (c) Oberlin, A., Tchoubar, C., Schiller, C. et
          Pézerat, H., Etude du fireclay produit par
          altération de la kaolinite et de quelques
          fireclays naturels. Colloque Inter. C.N.R.S.,
          No. 105, 45-57, 1961, C.N.R.S. - éd.

5.   (a) Mathieu-Sicaud, A., Mering, J. et Perrin-Bonnet, I.,
          Etude au microscope électronique de la mont-
          morillonite et de l'hectorite saturées par dif-
          férants cations. Bull. Soc. Franc. Minér. Crist.,
          74, 439-455, 1951

      (b) Mering, J., Oberlin, A. et Villière, J., Etude par
          électrodéposition de la morphologie des montmoril-
          lonites; Effet des cations calcium. Bull. Soc.
          Franc. Minér. Crist., 79, 515-522, 1956

      (c) Mering, J. et Oberlin, A., Vieillissement des sols
          d'hydroxydes de magnésium. Bull. Soc. Franc.
          Minér. Crist., 80, 158-165, 1957

6.   (a) Wyart, J., Oberlin, A. et Tchoubar, C., Etude de la
          boehmite formée par altération de l'albite.
          C.R. Ac. Sc., 256, 554-556, 1963

      (b) Tchoubar, C., Formation de kaolinite à partir
          d'albite altérée par l'eau à 200 $^{\circ}$C. Bull. Soc.
          Franc. Minér. Crist., 88, 483-518, 1965

7.   Magnan, C. Traité de Microscopie Electronique, Tome I,
          427-458, 1961. Hermann éd.

8.   Bates, Th.F. and Comer, J.J., Replica of clays.
                Clays and Clay Minerals, 3, 1-9, 1955

9.   Mitchell, B.D. and Mackenzie, R.C., Free iron oxide
                removal from clays. Soil Sci., 77, 173-184, 1954

10.   (a) Bates, Th.F., Selected Electron Micrographs of
                     Clays. Circular No. 1, College of Mineral
                     Industries, Penn. State Univ.,37-44, 1958

                (b) Hofmann, U., Scharrer, E., Czerch, W., Frühauf, K.
                     und Burck, W., Grundlagen der trocken gefrierten
                     Massen und die Ursachen der Raumerfüllung der
                     trockenen Masse. Ber.Deutsch. Keram. Ges.
                     39, 125-130, 1962

11.   (c) van Olphen, H. Clay Colloid Chemistry, 2nd Ed.
                     Wiley-Interscience

OECD RESULTS AND COMMENTS (Micrographs and patterns, figures 1-21)
    Rapporteur: A.Oberlin, C.N.R.S., Orléans, France.

01   Montmorillonite

The clay was brought in the sodium form by exchange with NaCl
and washing. Drops of an extremely dilute suspension (in the order
of a few ppm) were deposited, and the specimen was subsequently
shadowed. Micrograph 1 shows the typical morphology of very
thin sheets which are curled at the borders. The fine fraction (2)
consists of very thin particles, often showing angles of 120°
at the borders. The thickness of the particles is of the order of
10 A, hence, they consist perhaps of single layers. It is known
that such particles are composed of small domains with diameters of
the order of 100 A according to X-ray analysis. It is also known
that the symmetry group of the mineral as determined for the Wyoming
montmorillonite by SAD is C_{1m1}. One would,therefore, expect that
the diagram SAD for montmorillonites would only show hexagonal
symmetry in the intensities. However, the diagram of the thinnest
and best isolated particles (3) shows that this is not the case for
the Camp Berteau montmorillonite. This observation suggests that
the elementary domains in these particles are almost perfectly

mutually oriented at close to 60°, simulating a single crystal of
much larger diameter, resulting in complete hexagonal symmetry.
This is confirmed by dark field observations (4a,4b). When producing
the image by means of one of the intense beams (02) (11), one observes
that the particles brighten in a homogeneous fashion in the
coarse fraction, as well as in the fine fraction. The thicker particles
show numerous Moiré fringes, whereas the smallest ones show a uniform
brightness. This means, that the mutual orientation of the elementary
domains is so perfect that the particle consists only of a single
diffraction domain.

The oblique electron diffraction diagrams (5)is that of a two-
dimensional crystal. Only the reciprocal lines (20) and (13) are
modified by the variation of the structure factor when one moves
along Z in the reciprocal lattice.

02 Laponite

The particles are small and thin (6,7). Their diameter is between
200 and 300 A and their thickness is less than 20 A. The material is
very different from natural hectorite or saponite. The SAD patterns
(8) are composed of very diffuse bands which are characteristic of
turbostratic smectites.

03 Kaolinite (China Clay)

The major part of the sample consists of pseudohexagonal particles
(9) which are thin and of variable dimensions. The particles are very
rich in Moiré fringes and Bragg fringes, and numerous dislocations and
deformations can be discerned (micrograph 10 in bright field, and 11
in dark field-330). Each particle is monocrystalline (12). According to
SAD patterns the b-axis measures 8.96 A, which corresponds to that of
a kaolinite. That the kaolinite is particularly well crystallized is
confirmed by oblique electron diffraction patterns.

04 Attapulgite

The particles are typical elongated crystals (13), between 0.25 and
4 μm long and about 200 A in diameter. The fibers are rigid and are
easily broken which explains the variable length of the particles. They
tend to associate in a parallel fashion to form bundles, although less
pronounced as in chrysotile. The samples contain an impurity which
is probably a lamellar phyllite (13) , as well as quartz particles. The
SAD patterns show that the particles are elongated along the c-axis (14)
with a value c= 5.2 A.

05 Illite

The particles of the coarse fraction (15) have been examined by replicas of the powdered mineral. They consist primarily of mica and altered feldspar. The fine fraction (16) consists of small and irregular lamellar particles; their dimensions are smaller than 0.2 μm. SAD patterns are always taken of several particles together, hence they are of the Debye-Scherrer type. A b-value was derived of b = 9.05 ± 0.05 A. At the same time, one finds particles of impurities for which b = 9.2 ± 0.02 A.

06 Chrysotile

The tubular fibers (17) are very long and flexible. Their diameter varies between 200 and 500 A and there is a marked association of the fibers in bundles. All particles are elongated along the a-axis, with a value of a = 5.36 A. (18)

07 Crocidolite

The mineral is very pure, and the micrographs show well formed laths (19)

08 Talc

The fragments obtained by grinding are lamellar in shape with irregular contours (20)

10 Gibbsite

The replicas show that the particles often have hexagonal contours (21), and SAD patterns identify the mineral as gibbsite with c = 9.70 A.

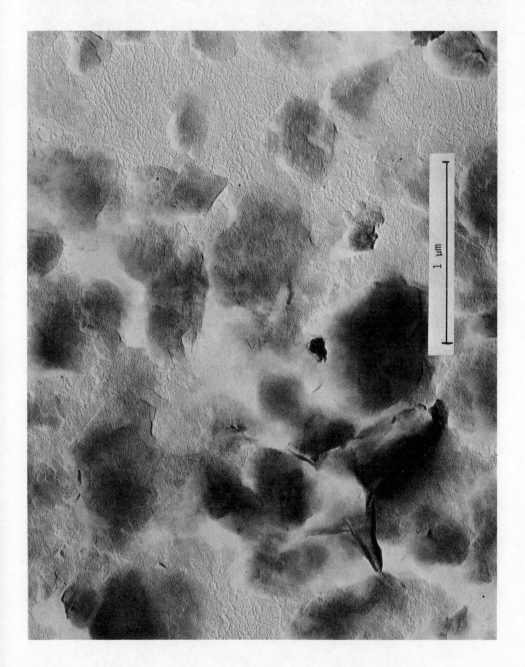

Figure 1   01 Montmorillonite.NiPd shadowed tan-1/3

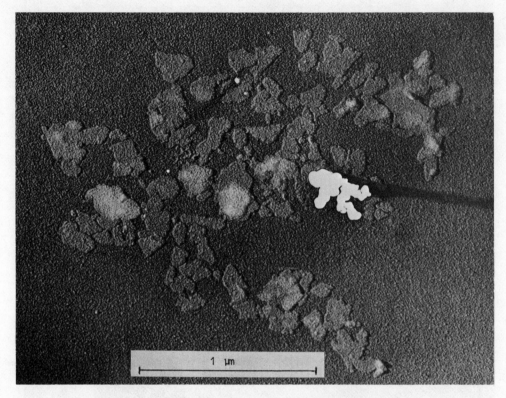

Figure 2   01 Montmorillonite.   Fine fraction.Cr shadowed α = 10°

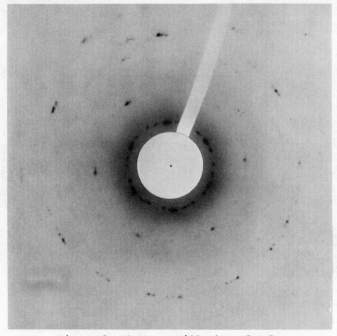

Figure 3   01 Montmorillonite . S.A.D.

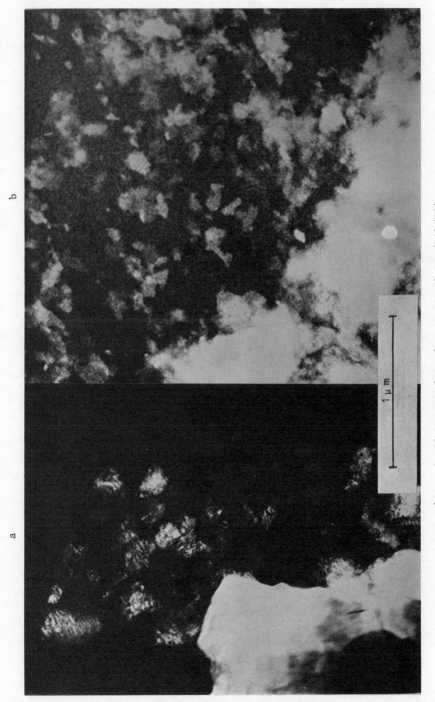

Figure 4   01 Montmorillonite. Dark field (02)(11)
a. coarse fraction
b. fine fraction

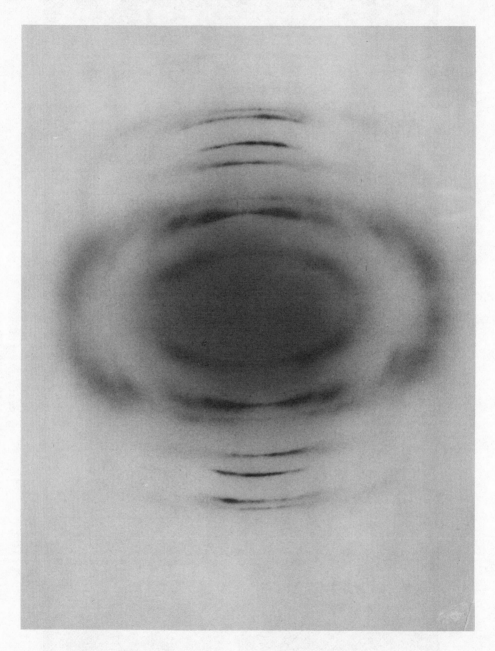

Figure 5   01 Montmorillonite. Oblique diffraction pattern at 60°

Figure 6   O2 Laponite. Cr shadowed $\alpha = 10°$

Figure 7   O2 Laponite. Double replica plastic-carbon. Shadowed WO_3 $\alpha = 30°$

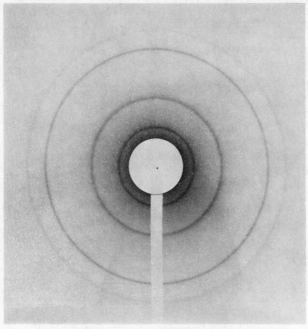

Figure 8   O2 Laponite. S.A.D.

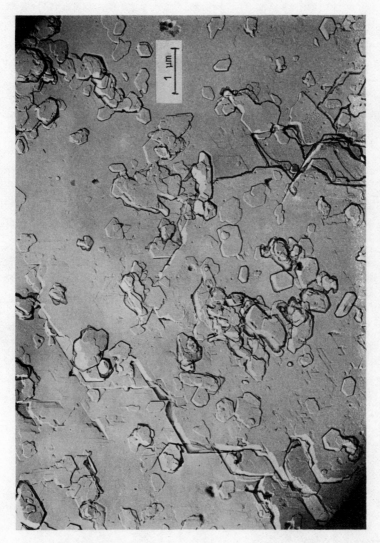

Figure 9   O3 Kaolinite. Double replica plastic–carbon. Shadowed WO_3 $\alpha = 30°$

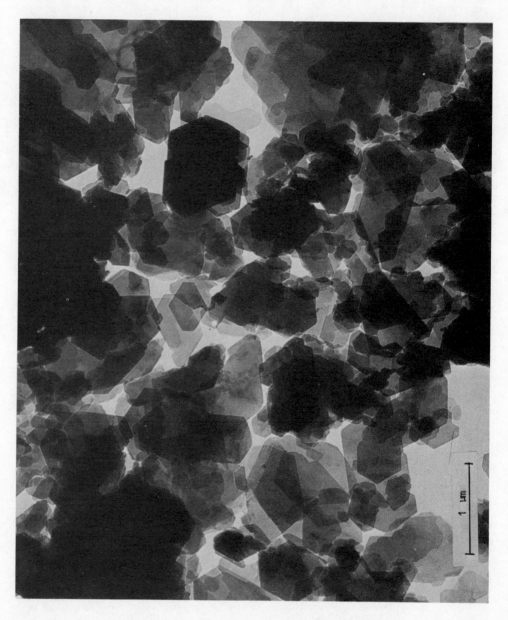

Figure 10 03 Kaolinite. Bright field

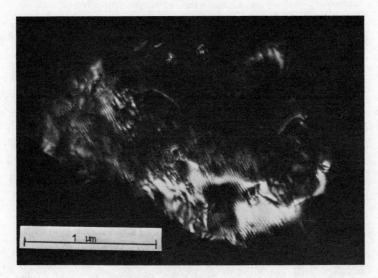

Figure 11 03 Kaolinite. Dark field

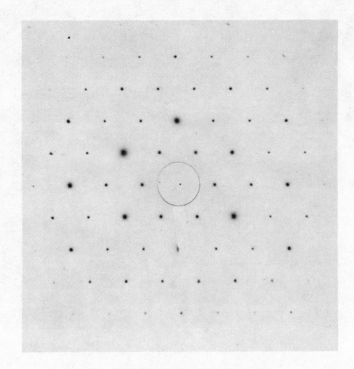

Figure 12   03 Kaolinite.S.A.D.

Figure 13   04 Attapulgite. Bright field

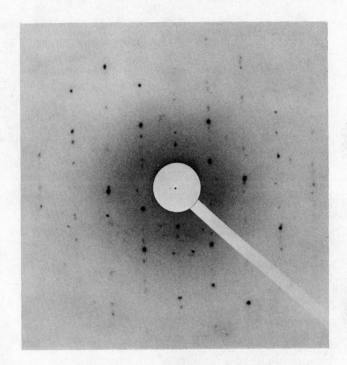

Figure 14   O4 Attapulgite.S.A.D.

Figure 15   O5 Illite.Coarse fraction pre-shadowed C-Pt tan-1/3. Direct replica

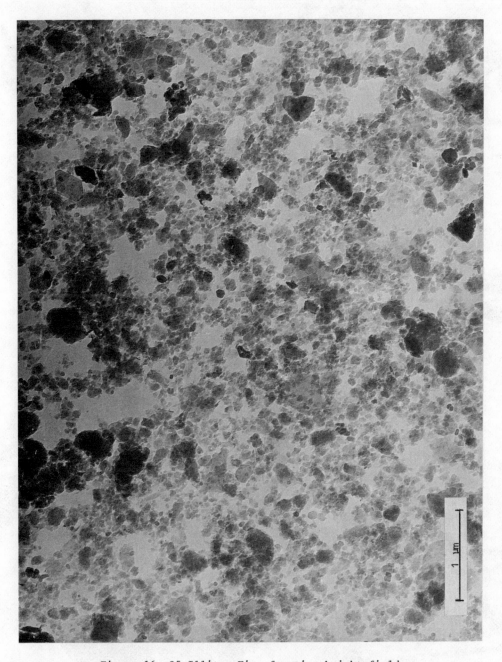

Figure 16 O5 Illite. Fine fraction bright field

Figure 17   06 Chrysotile.Bright field

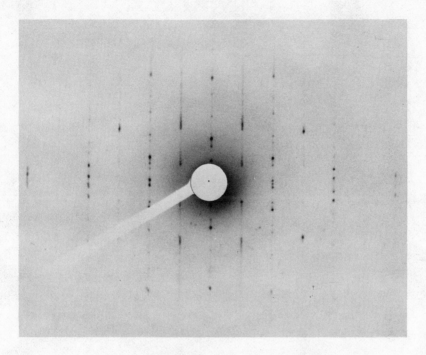

Figure 18   06 Chrysotile. S.A.D.

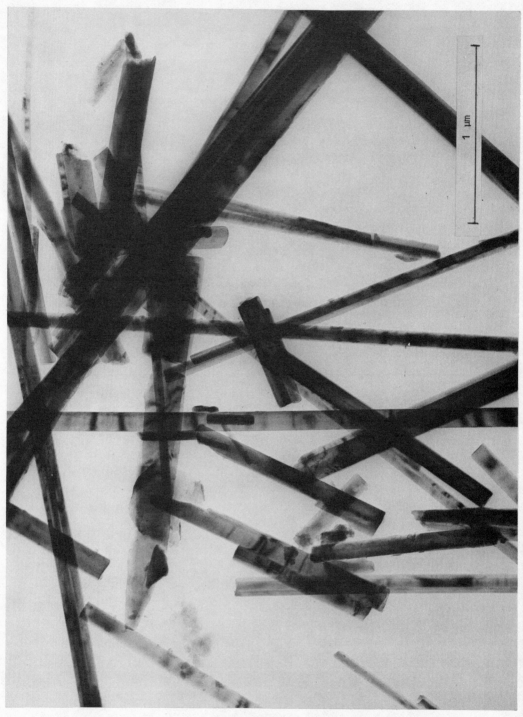

Figure 19 07 Crocidolite. Bright field

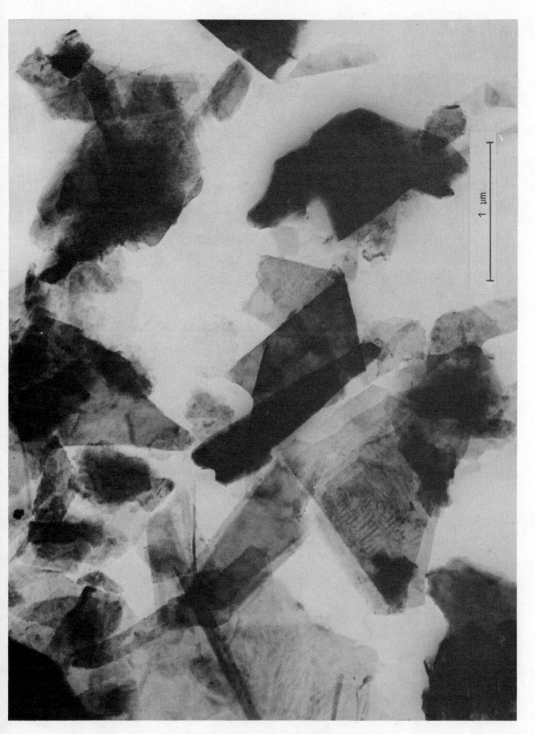

Figure 20 08 Talc. Bright field

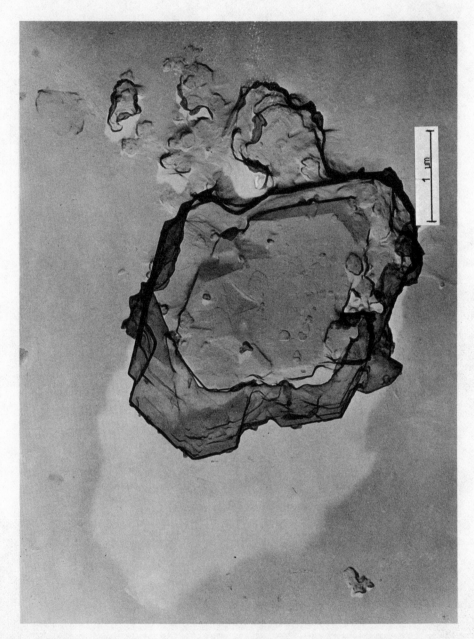

Figure 21 10 Gibbsite. Direct replica. Pre-shadowed C-Pt

THERMAL ANALYSIS, DTA, TG, DTG

R. C. Mackenzie and S. Caillère

INTRODUCTION

The three principal thermoanalytical techniques are differential
thermal analysis (DTA), thermogravimetry (TG) and derivative thermogravimetry (DTG).These terms were first adopted by the International,
Confederation for Thermal Analysis (Mackenzie, 1969) and subsequently
by IUPAC (1974). The significance of the results obtained by these
three methods is somewhat different, as described below.

(1) DTA

   DTA curves reveal all energy changes occurring in a sample on
heating. Such energy changes can arise from at least five causes: phase
transitions, decomposition, solid-state reactions in multi-component
samples, reactions with an active gas such as oxygen (generally surface
reactions) and second-order transitions (change in entropy without a
change in enthalpy). The curves are, therefore, a function of the
crystal sructure and chemical composition of the material(s) in the
sample and thus reflect the mineralogical nature of the sample
(Caillère and Hénin, 1947, 1948). In addition to use for "finger
printing", they can reveal the presence of small amounts of contaminants
and even differences in chemical composition or small differences in
structure. But DTA suffers from severe limitations (Mackenzie, 1970a,
1972; Garn, 1965) and can only give meaningful results if certain
practical precautions are observed (for a review of thermal analysis
see Mackenzie, 1974).

(2) TG and DTG

   TG and DTG curves reveal only weight changes occurring on
heating (due to decomposition or oxidation) and therefore yield more

limited information than DTA (Caillère, 1950). On the other hand, they
provide more accurate quantitative results, since many factors
associated with heat transfer through the samples (which are particularly
important in DTA) are inoperative. The DTG curve has some advantages
over the TG curve in that it clearly reveals minor changes in slope
that might well be missed on the original; moreover the temperatures
of the start and end of the peak on the DTG curve closely correspond
to the commencement, maximum rate and end of reaction – unlike the
DTA curve where these critical points have no or variable significance
(Mackenzie, 1970b) depending on the type of reaction and sample holder
configuration used. Peak temperatures on DTA and DTG curves are therefore
seldom, if ever, identical.

(3) General

Comparison of DTA and TG, or even better, DTG curves immediately
indicates which reactions relate to decomposition or oxidation. The
usual criterion for DTA is peak temperature but, since this is dependent
on many factors such as heating rate, sample size, particle size and
distribution, etc., a better reference point is the extrapolated onset
(Mackenzie et al.,1972; McAdie, 1972) which is near the commencement of
the reaction.

Pretreatment of samples

Samples should be of fine particle size – certainly less than
100-mesh, but care must be taken in trituration since grinding can
markedly affect crystal structure and hence the DTA curve (Mackenzie,
Meldau, and Farmer, 1956). Clay samples separated by sedimentation
(<2 μm) are already of satisfactory size, but if dried from suspension
must be rubbed gently (not ground) in an agate mortar to pass 100-
mesh. Fine grained samples in compact form may be dispersed by ultra-
sonic treatment without danger of alteration. Coarsely crystalline
samples present a problem and grinding may be unavoidable: if it is
absolutely necessary, it is recommended that grinding of such materials
should be as gentle as possible to reduce any deleterious effects to
the minimum.

For comparative thermal studies, it is recommended that all samples
with cation exchange properties be examined "as received" and also after
saturation with Ca^{2+}. It is also recommended that samples be equili
brated for at least four days in vacuo over a saturated solution of
$Mg(NO_3)_2.6H_2O$ (i.e. at 55% relative humidity) before carrying out
thermal investigations. This would ensure that differences due to
different humidity and temperature conditions are eliminated.

Apparatus and Techniques for DTA and TG

Many apparatus variables affect to varying degrees the nature of the curve obtained, particularly in DTA. Similarly, variations in technique can have a marked effect on the curves. Other aspects are covered by the code of practice for recording DTA and TG results, as

drawn up by the Standardization Committee of the International Conference on Thermal Analysis (McAdie, 1967). Thermocouples may be calibrated for temperature, and standards for this purpose are available from the U.S. National Bureau of Standards (McAdie, 1972)

Reporting of results

Recommendations for the reporting of results are the following: For DTA all extrapolated onset and peak temperatures (see Figure) should be marked to the nearest °C and, if possible, a scale provided on the ΔT axis indicating the deflection per °C at a specified temperature (Mackenzie et al., 1972).

For TG the extrapolated onset, maximum rate, and extrapolated completion temperatures should be given, and for DTG the procedural initial, the extrapolated onset and the peak temperatures. A weight scale should be provided on the abscissa for the TG curve and, if possible, a dw/dt or dw/dT scale for the DTG curve.

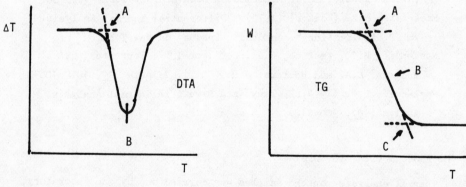

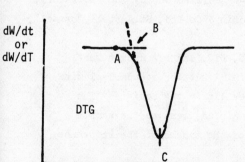

DTA : A = extrapolated onset
      B = peak location

TG : A = extrapolated onset
      B = maximum rate
      C = extrapolated completion

DTG : A = procedural initial
      B = extrapolated onset
      C = peak location

References

Caillère,S. and Hénin, S. (1947). Ann. Agron., No.1, pp. 1-50

Caillère, S.and Hénin, S. (1948). Verres Silic. Ind., 13, 114-118

Caillère, S. (1950). Trans.Fourth Int.Congr.Soil Sci.,Amsterdam,
    3, 54-62

Garn, P.D. (1965). "Thermoanalytical Methods of Investigation",
    Academic, New York.

IUPAC (1974) Pure and Applied Chemistry, 37, 441-444

Lombardi,G.(1977). "For Better Thermal Analysis"
    ICTA and Instituto di Mineralogia e Petrografia dell'Università
    di Roma, Rome.

McAdie, H.G. (1967). Analyt.Chem., 39, 543

McAdie, H.G. (1972). In "Thermal Analysis : Proc. Third ICTA, Davos,
    1971" (H.G.Wiedemann, Ed.). Birkhäuser Verlag, 1, 591-608

Mackenzie, R.C., Meldau, R. and Farmer, V.C. (1956). Ber. Deut.
    Keram.Ges., 33, 222-229.

Mackenzie,R.C.(1969) Talanta,16, 1227

Mackenzie, R.C. (Editor) (1970a). "Differential Thermal Analysis -
    Vol. 1 . Fundamental Aspects".Academic, London and New York

Mackenzie, R.C. (1970b). In "Differential Thermal Analysis"
    (R.C.Mackenzie, Ed.). Academic, London and New York, 1, 3-30

Mackenzie, R.C. (Editor) (1972). "Differential Thermal Analysis.
    Vol.2. Applications". Academic, London and New York.

Mackenzie, R.C., Keattch, C.J., Dollimore, D., Forrester, J.A.,
    Hodgson, A.A. and Redfern, J.P. (1972). Talanta, 19, 1079-1081

Mackenzie, R.C. (1974).Highways and Byways in Thermal Analysis,
    Analyst, 99, 900-912.

CMS RESULTS (DTA, TG, DTG)

Thermal analysis on the samples was carried out by one laboratory,
J.B.Rowse, English Clays Lovering Pochin & Co.Ltd, St.Austell,Cornwall,
U.K.

Using a Mettler Thermoanalyzer, TG, DTG, and DTA curves were
recorded at a heating rate of 10°C/min. A Mettler standard alumina
block was used with platinum crucibles lining the cavities. The
recorded temperature was measured in air adjacent to the sample
holder - it was not the temperature of the sample or the reference
material.

CMS: DTA AND TG

	sample weight mg	ads. water %	struct. water %	DTA Peak temperatures °C
KGa-1 Kaolinite, well crystallized	424.6	0.26	13.11	630- §, 1015+
KGa-2 Kaolinite, poorly crystallized	363.2	0.74	13.14	625- §, 1005+
SWy-1 Montmorillonite Wyoming	414.8	4.99	5.53	185- °, 235- ° (sh), 755- §, 810- (sh), 980+
STx-1 Montmorillonite Texas	358.9	12.45	3.88	185- °, 240- ° (sh), 720- §, 920-, 1055+, 1090+, 1135+
SAz-1 Montmorillonite Arizona	403.5	12.76	4.69	200- °, 240- ° (sh), 685- §, 895-, 1020+, 1065+, 1160+
SHCa-1 Hectorite California	469.3	3.34	20.28°°	165- °, 725 (sh), 795- §, 880-, 910-(sh), 1130-
Syn-1 Synthetic mica-montmorillonite	230.4	3.56	10.35	140- °, 575- §, 1030+
PFl-1 Attapulgite Florida	239.2	12.96	5.52	170- °, 230-→300- (sh) °, 495-, 550- §, 840-, 905+

(°) desorption of water; (§) dehydroxylation; (°°) includes CO_2 from calcite impurity
- endothermic peaks; + exothermic peaks; (sh) shoulder.

% adsorbed water based on original sample weight; % structural water based on sample weight
at a point just prior to dehydroxylation on the TG curve.

The DTA curve was recorded using Pt-Pt,10% Rh thermocouples
situated at the centers of the specimen and reference materials.
The specimens were packed into the crucible with a plunger using
moderate hand pressure. The bulk densities of the various materials
led to variations in the weights taken, resulting in slight variations
in the peak temperatures.

The results are shown in the graphical records, and the data
are summarized in a table.

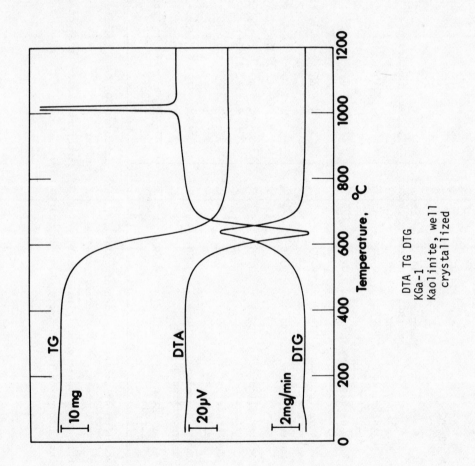

DTA TG DTG
KGa-1
Kaolinite, well
crystallized

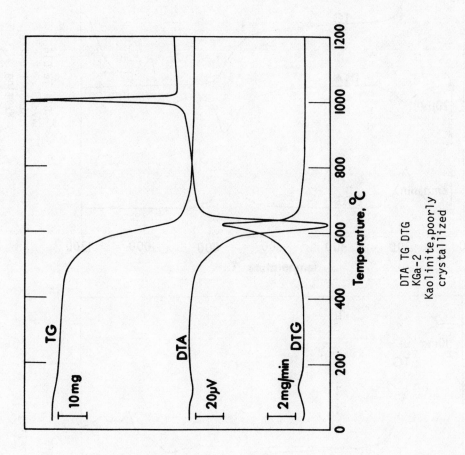

DTA TG DTG
KGa-2
Kaolinite, poorly
crystallized

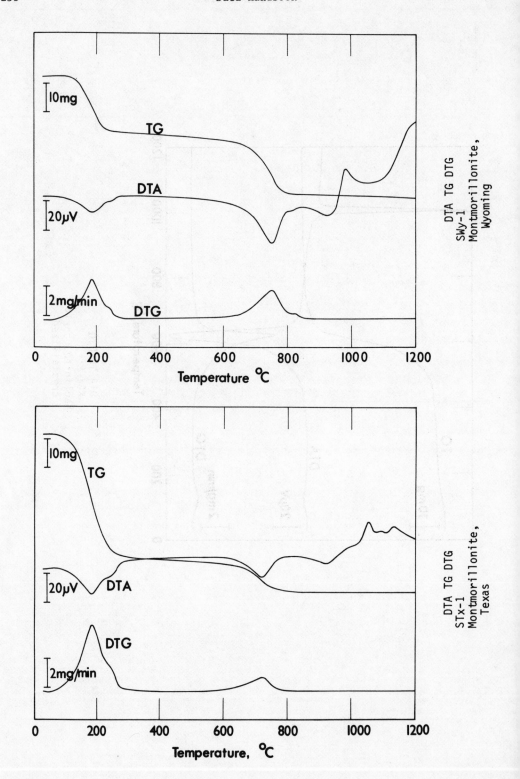

DTA TG DTG
SWy-1
Montmorillonite,
Wyoming

DTA TG DTG
STx-1
Montmorillonite,
Texas

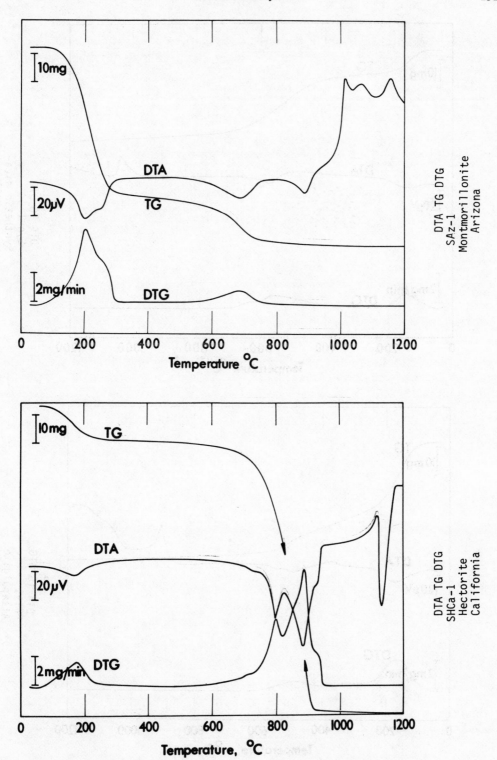

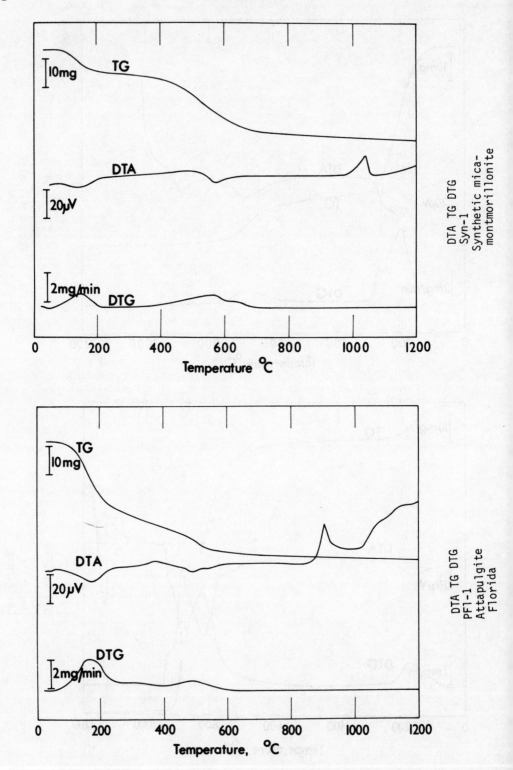

DTA TG DTG
Syn-1
Synthetic mica-
montmorillonite

DTA TG DTG
PF1-1
Attapulgite
Florida

CMS COMMENTS (DTA, TG, DTG)

DTA curves are representative of the principal mineral constituents. Except for hectorite, impurities are not present in sufficient quantity to show well defined peaks on the DTA or DTA curves, but they may account for some of the shoulders seen on the main peaks.

The hectorite sample contains a considerable amount of calcite which gives rise to the endothermic peak at 880°C due to the evolution of CO_2, and this peak shows the sharp termination characteristic of calcium carbonate decomposition. There is overlap of the dehydroxylation and release of CO_2, hence it is not possible to estimate accurately the percentage of calcite in the hectorite sample. However, if the weight loss above 810°C (the minimum in the DTG curve) is allocated to calcite, there is approximatily 27% of this mineral present in the hectorite. The endotherm at 1130°C probably is the result of the reaction between CaO and the dehydroxylated hectorite. There was also sintering of the sample.

Low temperature losses of adsorbed water of the smectites vary depending on the storage conditions prior to the determination. The presented thermal analysis curves were recorded on the minerals "as received".

Dehydroxylation losses for kaolinite are close to the theoretical value of 14%, and indicate that less than 7% impurities are present.

Dehydroxylation losses for the montmorillonites from Arizona and from Texas are somewhat lower than the theoretical value of about 5%. It is not possible to give an exact value because of overlap of loss of adsorbed water and loss of hydroxyl groups, in spite of the large temperature difference between the two events. The montmorillonite from Wyoming shows a loss of hydroxyl groups which is slightly greater than the theoretical value. The high value for the weight loss in the dehydroxylation range for the synthetic mica-montmorillonite is probably due to the loss of ammonia together with water from hydroxyl groups since this clay contains ammonium in exchange positions (see under C.E.C.). The DTG curve shows a shoulder which may well arise from the loss of a second volatile component.

Where there is a weight loss, the DTG curve mirrors the DTA curve in almost every case.

OECD RESULTS (DTA)

The number of laboratories submitting results on individual samples varied between 2 and 14. Generally, a heating rate of 10°C/min. was used. The general shape of the DTA curves is represented by the schematic figures, and the table lists the ranges of observed peak temperatures, the mean peak temperatures, and the range in which most reported peak temperatures were observed.

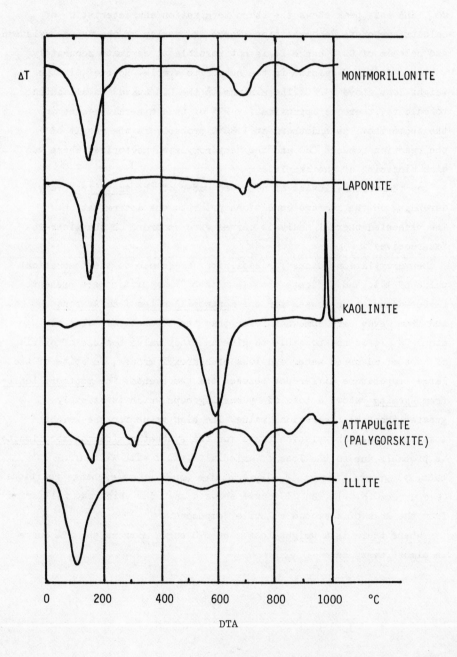

DTA

OECD: DTA

Ranges and means of peak temperatures obtained at a heating rate of approximately 10°C/min.

	§	peak temperatures °C max. range	mean	most results	other peaks	
01 Montmorillonite	14	668–713	691	690–700	150	aw
		830–901	867	860–880	560	i
		915–995+	949+	940–950+		
02 Laponite	7	675–717	690	680–710	130	aw
		695–738+	709+	710–720+		
03 Kaolinite (China Clay)	11	567–610	592	600–610	80	aw
		960–995+	978+	980–990+		
04 Attapulgite (Palygorskite)	10	273–320	298	300–310	160	aw
		477–510	495	500–510	573	i
		905–940+	917+	910–920+	740	i
05 Illite	9	530–580	562	570–580	100	aw
		856–902	875	870–880		
06 Chrysotile	4	690–705	699	690–700	100	aw
		800–817+	809+	800–810+		
07 Crocidolite	2	430–440+	434+	430–440	110	aw
		868–929	899		730	i
08 Talc	6	955–983	970	960–980	100	aw
					570	i
					620	i
					730	i
					870+	i
10 Gibbsite	6	230–270	247	230–250		
		308–360	327	310–320		
		527–540	533	530–540		
11 Magnesite	7	625–695	661	650–690	50	aw
					573	i
					770	i
12 Calcite	7	895–935	915	900–930		
13 Gypsum	6	136–197	171	160–190		
		180–222	203	180–210		
		363–390+	374+	370–380+	800	i

---

§ : number of laboratories submitting results
+ : exothermic reaction
aw: peak due to desorption of adsorbed water
i : very small peak or shoulder due to impurities

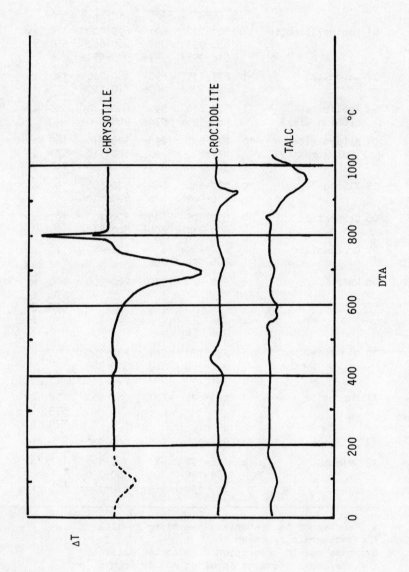

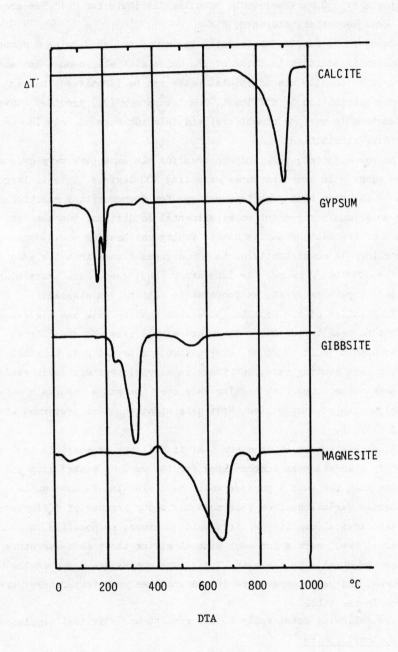

DTA

OECD COMMENTS (DTA)

Rapporteur: R.C.Mackenzie, The Macaulay Institute for Soil Research,
    Craigiebuckler, Aberdeen, U.K.

Some of the samples are relatively impure and show quite a number
of peaks in addition to those of the major mineral. In most instances
impurities causing the additional peaks can be identified, but in
others attribution is difficult. One laboratory (I1) provided curves
determined in various atmospheres and this has been of help in
deciding attribution.

As was expected, peak temperatures for the same peak vary over a
wide range - in some instances exceeding 100 degrees. This is largely
due to the fact that peak temperatures for decomposition reactions
are notoriously dependent on experimental conditions, but some at
least of the scatter may be due to faulty calibration - e.g. one
laboratory is consistently up to 100 degrees higher than the mean
in the 900-1000°C range. One laboratory (GR3) used phase transition
points of pure materials as temperature calibration standards.

Examination of the results shows that heating rate has a greater
effect on peak temperatures than does sample size. No consistent
relationship could be found between sample size and peak temperature
at the same heating rate, but there is a consistent marked increase
in peak temperature with heating rate over the range 3-20°/min
irrespective of sample size. Most determinations were performed at
10 ± 1 °/min.

From this survey it is clear that it will not be possible to
specify standard peak temperatures for the various peaks: adoption
of the mean for such a purpose would be misleading because quite
authentic variations from this mean may occur because of differences
in experimental conditions. It should, however, be possible to
produce "type" curves for each mineral giving the peak temperature
range, together with the mean. These type curves are shown in the
figures, and peak temperature ranges and mean peak temperatures are
given in the table.

The following notes apply to the results on individual samples:

01 Montmorillonite

A fairly typical curve for montmorillonite; the very small peak on
most curves at about 560°C could be due to a small amount of contaminating kaolinite and perhaps also quartz. One laboratory (D28) reports
calcite in the >20 μm fraction, but no peak due to calcite can be

detected on any of the other curves, the amount of calcite in the
original material must therefore be very small. Impurities in general
do not much affect the curves.

02 Laponite

This material shows a large hygroscopic moisture peak followed by an
exothermic-superposed-on-endothermic system at about 700°C. The curve
obtained by one laboratory (F1) differs from all the others in that
the exothermic component is very poorly developed.

03 Kaolinite (China Clay)

Typical kaolinite curve suggestive of highly crystalline kaolinite,
but with a small hygroscopic moisture peak. Any impurities present
(mica is mentioned) do not affect the curve.

04 Attapulgite

This sample contains quartz and a small amount of calcite, both of
which show up on all DTA curves obtained at sufficient sensitivity
(peaks at 573 and ca 740°C). The doubling of the peak at 300°C suggests
the presence of gibbsite or goethite. Otherwise, the curve is typical
for palygorskite.

05 Illite

Curves generally above 300°C are fairly typical for illite, but
the hygroscopic moisture peak is very much larger than normally
shown by this mineral, suggesting the presence of non-crystalline
material or interstratified montmorillonite layers. The small amount
of quartz reported by one laboratory (D25) does not affect the curve
at the sensitivity used.

06 Chrysotile

The sample is apparently 98% pure (see TG results) and impurities
do not seem to affect the curves at normal sensitivities. The DTA
curve is typical, but one laboratory (D25) reports a doubling of the
exothermic peak at about 800°C.

07 Crocidolite

Curves for this material were obtained at rather low sensitivities,
and characteristic effects are, therefore, not well defined. One
laboratory (I1) employed various atmospheres (air, argon and CO_2)
and the results clearly show that the endothermic peak at about 730°C
is probably due to contaminating calcite, and that the exothermic
peak at 434°C is due to an oxidation reaction. The main endothermic
event is at about 900°C. Contaminants do not affect the curves at
the sensitivities used.

08 Talc

This is a very impure sample containing chlorite and, according to one laboratory(D25) quartz. In addition to the talc peak at about 950°C (dehydroxylation) there is a double endothermic peak at about 570-620°C, another endotherm at about 730°C, and a very small exotherm at about 870°C. The 870°C exotherm belongs to chlorite, but the attribution of the other peaks is not quite certain, although one of the endotherms certainly belongs to chlorite.

10 Gibbsite

The curve is typical of this mineral which seems to be pure. The peaks at 251 and at 329°C are due respectively to the formation and the dehydroxylation of boehmite resulting from the thermal treatment.

11 Magnesite

This sample is impure, and according to two laboratories (D25 and GR3) contains 5-10% dolomite (double endothermic peak at 770°C) and some quartz (weak endothermic peak at 573°C). The dolomite peak shows up on all curves, and quartz appears where sensitivity is adequate. Smaller amounts of calcite, talc, and possibly chlorite have been detected by GB25. Most DTA curves show an exothermic peak before the main decarbonation peak, but others after.

12 Calcite

This sample yields a typical curve and would appear to be relatively pure.

13 Gypsum

The sample shows a reasonably characteristic curve but seems to contain some calcite (or dolomite?) yielding a peak or double peak in the 750-820°C region. The peaks at 171 and 203°C are due to respectively the formation of the hemihydrate and the anhydride.

OECD RESULTS (TG)

Thermogravimetry, mostly along with DTA, was carried out by 19 laboratories. The largest number of determinations on a single sample (about 8) was reported for the silicates.

Adamel and Stanton equipment was used most frequently, but other thermobalances (Ugine Reynaud, Netzsch, Cahn, and Shimadzu) were employed by some. Heating rates varied between 100°C/h and 650°C/h (1.66 and 11 °C/min) but most laboratories used rates between 150°C/h and 200°C/h (2.5 and 3.33 °C/min). Sample weights were mostly in the 100-250 mg range,and between 2 and 10 mg respectively, for the Cahn and Shimadzu microbalances.

The results are summarized in a series of tables, and for most of
the samples one of the submitted curves is presented in order to show
their general shape. (Contrary to present recommended practice, the
weight loss, rather than the sample weight, is plotted on the ordinate).

OECD THERMOGRAVIMETRY

01   Montmorillonite

Lab. Equipment Conditions	Treatment of sample			DTA 692	DTA 881	DTA+ 950		§ corr.:
F2 Ugine Reynaud air	none 100 mg	20-225° 15.25%	225-530° 1.65%	530-710° 3.30%		710-1000° 0.90%	20-530° 16.90%	530-710° 3.95%
F3 Adamel air 150°/h	none 250 mg	45-175° 12.50%	175-480° 1.80%	480-695° 3.20%			45-480° 14.30%	480-695° 3.75%
	eq.55% R.H. 250 mg	40-175° 10.0%	175-500° 2.00%	500-695° 3.00%			40-500° 12.00%	500-695° 3.40%
F6 Adamel air 150°/h	none 250 mg	20-190° 15.30%	190-500° 1.40%	500-700° 3.30%		700-1000° 0.60%	20-500° 16.70%	500-700° 3.90%
	350 mesh fr. 250 mg	20-190° 14.80%	190-500° 1.20%	500-700° 3.35%		700-1000° 0.65%	20-500° 16.00%	500-700° 3.60%
F9 Adamel air 150°/h	none 400 mg	20-180° 18.10%	180-500° 1.90%	500-700° 3.30%		700-1000° 0.30%	20-500° 20.00%	500-700° 4.10%
F16 Adamel N₂ 100°/h	eq.55% R.H. 400 mg	20-160° 14.15%	160-555° 1.70%	555-670° 3.15%		670-1000° 0.95%	20-555° 15.85%	555-670° 3.75%
D5 Stanton air 325°/h	none 100 mg	20-180° 13.60%	180-540° 1.10%	540-750° 3.20%			20-540° 14.70%	540-750° 3.75%
P3 Stanton air	none 1.0 g	25-250° 15.40%	250-500° 1.05%	500-700° 3.00%		700-1000° 0.50%	25-500° 16.45%	500-700° 3.60%
	Ca form	25-250° 17.10%	250-500° 1.00%	500-700° 3.10%		700-1000° 0.50%	25-500° 18.10%	500-700° 3.80%

§ corr.: calculated on dry weight (original weight – weight loss below this temperature range)

02   Laponite

Lab.	Equipment Conditions	Treatment of sample		DTA 696⁻	DTA 717⁺		
F3	Adamel air 150°/h	none	20-175° 8.35%	175-640° 3.55%	640-720° 2.40%		175-720° 5.95%
F5		none	20-140° 10.50%	140-680° 3.40%	680-720° 2.40%	720-1000° 0.20%	140-720° 5.80%
F9	Adamel air 150°/h	none 400 mg	20-100° 22.50%	100-620° 2.50%	620-700° 2.50%		100-700° 5.00%
F16	ICF N₂ 100°/h	55% R.H. 400 mg	20-200° 12.00%	200-620° 3.50%	620-720° 2.30%	720-1000° 0.50%	200-720° 5.80%
	air 100°/h	none 2.0 g	18.00%	3.50%	2.10%	0.10%	5.60%
D5	Stanton N₂ 325°/h	55% R.H. Ca form 100 mg	20-200° 17.00%	200-600° 2.80%	600-800° 2.00%		200-800° 4.80%
JAP3	Shimadzu N₂ 600°/h	none 218.2 mg	20-200° 20.00%	200-650° 3.50%	650-800° 2.00%		200-800° 5.50%

O3   Kaolinite (China Clay)

Lab.	Equipment Conditions	Treatment of sample	DTA-595	DTA-978+	§
F3	Adamel air 150°/h.	none	480-645 10.50%		10.50%
F5	Adamel air 150°/h	none	480-570° 12.60%		12.60%
F6	Adamel air 150°/h	none 250 mg	460-800° 12.65%		12.65%
F9	Adamel air 150°/h	none 400 mg	400-850° 12.60%		12.60%
F16	ICF N₂ 100°/h air	55% R.H. 400 mg	20-460° 0.35% 460-610° 11.95%	610-1000° 0.30%	12.25%
		2.0 g	20-470° 0.90% 470-570° 12.00%	570-1000° 0.60%	12.60%
F19	Adamel air 150°/h	55% R.H. 496.7 mg	420-575° 12.50%		12.50%
D5	Stanton air 325°/h	55% R.H. Ca form 100 mg	440-720° 11.00%		11.00%
P3	Stanton	none 100 mg	25-350° 0.65% 350-650° 11.85%	650-850° 0.90% 850-1000° 0.15%	12.90%
		Ca form 100 mg	0.65% 11.90%	0.80% 0.15%	12.85%
GB24	Stanton	none 225.4 mg	480- 12.95%		12.95%
GB37	Stanton air 240°/h	none 440 mg			12.85%

§ total, less low temperature loss

O4   Attapulgite

Lab.	Equipment Conditions	Treatment of sample		DTA 299 -	DTA 496 -	DTA 834 - DTA 921 +	total
F1		none	20-180° 7.05%	180-310° 2.55%	310-570° 4.15%	570-1000° 1.65%	15.40%
		600 mg					
F2	Ugine Reynaud air	none	20-165° 6.95%	165-290° 2.55%	290-585° 4.35%	585-725° 1.75%	16.35%
		100 mg					
F5	Adamel air 150°/h	none	20-215° 7.90%	215-390° 2.40%	390-630° 4.20%	630-690° 1.50%	16.00%
F6	Adamel air 150°/h	none	20-200° 7.20%	200-375° 2.60%	375-600° 3.80%	600-1000° 1.60%	15.20%
		250 mg					
F9	Adamel air 150°/h	none	20-115° 8.40%	115-225° 2.50%	225-540° 4.20%	540-640° 1.40%	16.50%
		400 mg					
F16	ICF N$_2$ 100°/h	55 % R.H.	20-140° 5.40%	140-250° 2.70%	250-495° 3.70%	495-1000° 2.55%	14.35%
		400 mg					
F19	Adamel air 150°/h	none	50-195° 6.10%	195-240° 2.80%	240-460° 5.70%	460-700° 2.90%	17.50%
GB37	Stanton air 240°/h	none	20-250° 7.50%	250-350° 2.50%	350-550° 3.80%	550-750° 2.30%	16.10%
		472 mg					
F16	ICF air 100°/h	55 % R.H.	20-130° 6.20%	130-240° 2.80%	240-480° 3.80%	480-1000° 2.40%	15.20%
		2.0 g					

05   Illite

Lab.	Equipment Conditions	Treatment of sample			DTA_565	DTA_878	DTA_927+		corr. §
F5	Adamel air 150°/h	none	20-100° 15.00%	100-410° 0.50%	410-580° 2.90%		580-1000° 0.80%	100-580° 3.50%	4.10%
F6	Adamel air 150°/h.250 mg	none	20-150° 5.30%	150-400° 1.00%	400-600° 2.65%		600-1000° 1.00%	150-600° 3.65%	3.85%
F16	ICF N₂ air	55 % R.H. 400 mg	20-120° 7.90%	120-400° 0.50%	400-590° 3.10%		590-1000° 0.80%	120-590° 3.60%	3.90%
	air	55 % R.H. 2.0 g	20-130° 9.80%	130-420° 1.20%	420-580° 2.80%		580-1000° 0.30%	130-580° 4.00%	4.40%
F19	Adamel air 150°/h	fr.<25 µm 495.2 mg	40-220° 6.75%		350-700° 3.65%			350-700° 3.65%	3.90%

§ corr. : calculated on dry weight (original weight – weight loss first column)

06 Chrysotile

Lab.	Equipment Conditions	Treatment of sample			DTA 697⁻	DTA 809⁺
F9	Adamel air 150°/h	none 400 mg	30-200° 3.60%	200-475° 0.75%	475-750° 11.50%	750-1000° 0.55%
GB36	Stanton N₂ 175°/h	none 500.7 mg		20-525° 1.95%	525-725° 11.50%	725-900° 0.60%
GB37	Stanton air 240°/h	none 469 mg				-1000° 13.55%

07 Crocidolite

Lab.	Equipment Conditions	Treatment of sample				DTA 430⁺	DTA 926⁻	
I1	Cahn RG Ar 180°/h	none 11.19 mg	20-100° 0.20%	100-350° 0.60%	350-410° +0.10%	410-650° 1.75%	650-950° +0.70%	950-1000° 0.40%
GB7	Stanton N₂ 180°/h	none 146 mg			20-410° 0.80%	410-535° 1.50%		

08   Talc

Lab.	Equipment Conditions	Treatment of sample				DTA-974	total
I1	Cahn RG Ar 190°/h	none 8.80 mg	20–450° 0.60%	450–650° 1.10%	650–795° 0.30%	795–1000° 4.25%	6.25%
F1	Adamel air 150°/h	none 1.0 g	20–530° 0.35%	530–695° 0.70%		695–1000° 4.20%	5.25%
F3	Adamel air 150°/h	none	20–600° 0	600–680° 0.25%	680–880° 0.25%	880–990° 3.70%	4.20%
F5	Adamel air 150°/h	none	20–550° 0.40%	550–580° 0.15%	580–870° 0.55%	870–990° 4.20%	5.30%
F6	Adamel air 150°/h	none 250 mg			20–875° 1.40%	875–1000° 3.80%	5.20%
F9	Adamel air 150°/h	none 400 mg			20–800° 1.20%	800–1050° 4.35%	5.55%
F16	ICF N₂ 100°/h air	none 400 mg 2.0 g	20–550° 0.10% 0.70%		550–850° 1.50% 0.80%	850–1000° 4.05% 4.20%	5.65% 5.70%
GB37	Stanton air 240°/h	none 629 mg			20–750° 1.20%	750–1000° 4.35%	5.55%

Lab.	Equipment Conditions	Treatment of sample		DTA 251-	DTA 659-	DTA 329-	DTA 540-	total
10	**Gibbsite**							
B5	Adamel air 150°/h	none	20-200° 0.35%	200-265° 7.80%	265-300° 18.00%	300-470° 3.75%	470-550° 4.10%	34.00%
GB7	Stanton N_2 180°/h	none 205.3 mg						32.90%
GB37	Stanton air 240°/h	none 603 mg			20-325° 26.00%		325-550° 8.30%	34.30%
11	**Magnesite**							
F5	Adamel air 150°/h	none	20-515° 1.61%	515-600° 40.60%	600-640° 0.90%	640-690° 2.70%	690-1000° 0.70%	46.50%
F16	ICF N_2 100°/h	55 % R.H. 100 mg	20-595° 1.30%	595-630° 41.50%	630-700° 4.10%		700-1000° 1.90%	48.80%
F16	air		20-490° 0.60%	490-620° 42.80%	620-720° 3.40%		720-1000° 0.20%	47.00%
EIR1	Netzsch N_2 120°/h	none 317.5 mg		20-630° 43.50%			630-1000° 3.75%	47.25%
I1	Cahn RG Ar 180°/h	none 10.61 mg	20-450° 2.05%	450-650° 43.25%	650-700° 0.85%		700-1000° 0.30%	46.45%
JAP3	Shimadzu air 600°/h	none 203.1 mg		25-730° 42.50%	730-850° 2.50%		850-1000° 0.25%	45.25%
B3	Adamel	none 400 mg						44.50%
GB7	Stanton	55 % R.H. 175 mg						46.80%
GB25	Microbalance	none 2.02 mg						47.00%
GB37	Stanton	510 mg						43.65%

12   Calcite

Lab.	Equipment Conditions	Treatment of sample			DTA 925⁻		total
F1	Adamel air 150°/h	none 250 mg	20-500° 1.70%		500-1000° 42.40%		44.10%
F2	Ugine Reynaud air	none 50 mg	20-545° 0.50%		545-870° 43.80%	870-1000° 0	44.30%
F3	Adamel air 150°/h	none	20-490° 1.10%		490-790° 44.70%		45.80%
F16	ICF N$_2$ 100°/h	none 100 mg	20-500° 0.80%	510-570° 0.20%	570-790° 42.50%	790-1000° 0.50%	44.00%
	air	2.0 g	20-460° 0.40%	460-650° 0.05%	650-840° 43.95%	840-1000° 0.05%	44.45%
B3	Adamel	none					43.10%
EIR1	Netzsch	none 300 mg					43.30%
GB7	Stanton	none 266 mg					44.75%
GB37	Stanton	none 404 mg					44.20%

13   Gypsum

Lab.	Equipment Conditions	Treatment of samples	DTA 163⁻	DTA 193⁻	DTA 374⁺			total
F5	Adamel air 150°/h	none	20-120° 19.50%	120-640° 0.25%	640-710° 2.40%	710-1000° 0.15%		22.30%
B3	Adamel air 150°/h	none 400 mg	20-230° 19.00%	230-500° 0.30%	500-900° 2.50%			21.80%
GB7	Stanton N₂ 180°/h	55 % R.H. 198.2 mg	20-120° 19.10%	120-670° 2.00				21.10%
EIR1	Stanton N₂ 180°/h	none 162,8 mg	20-120° 18.50%	120-670° 2.00%				20.50%

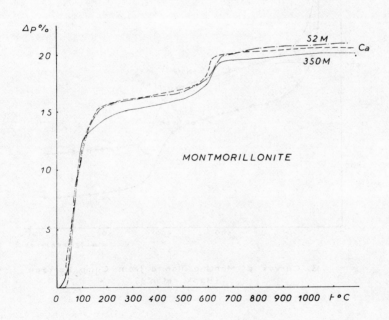

MONTMORILLONITE

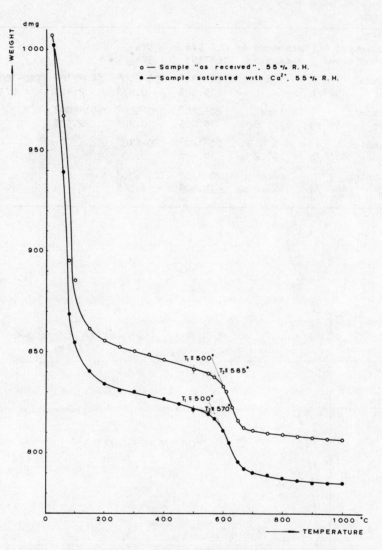

TG Curves of Montmorillonite from Camp Berteau.
(Bank ref. n° 1)

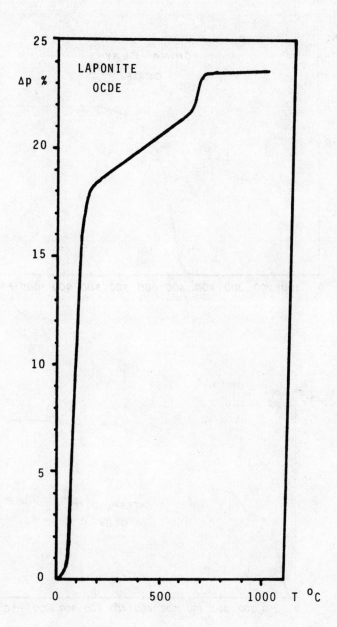

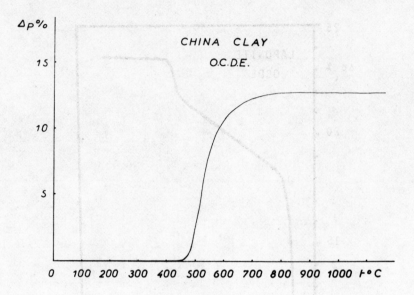

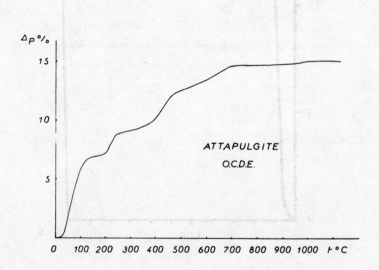

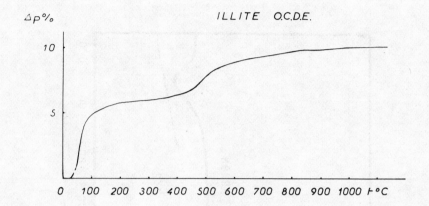

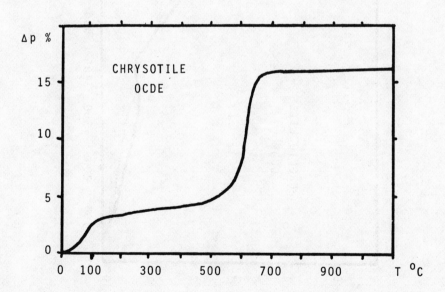

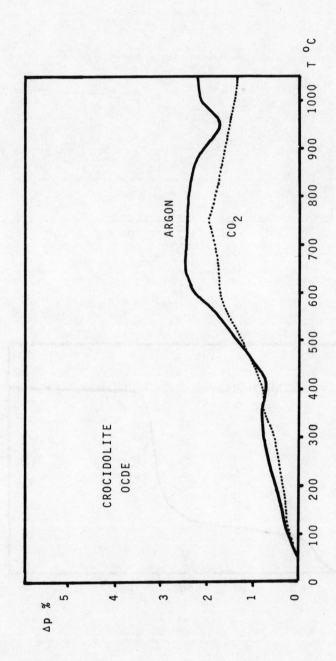

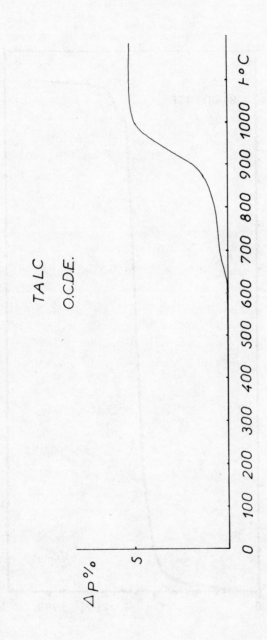

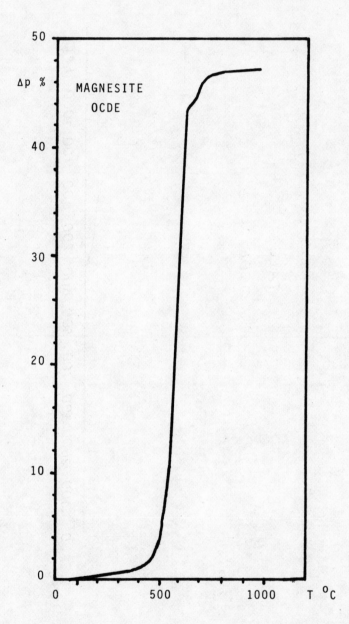

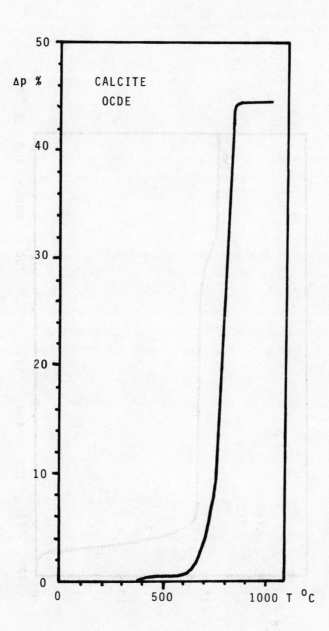

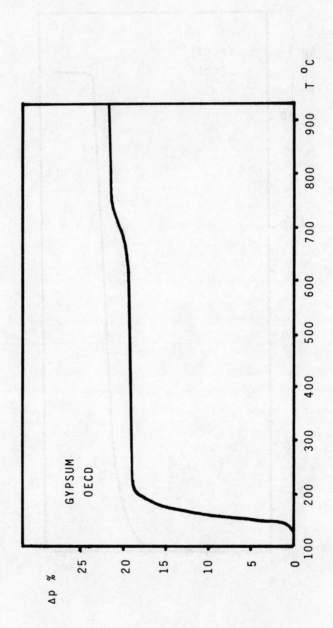

OECD COMMENTS (TG)

Rapporteur: S.Caillère, Muséum National d'Histoire Naturelle, Paris.

The data presented in the tables were either read from the records
by the authors or by us.

Temperature ranges are selected in which an important event takes
place as indicated by both the TG and the DTA curves. Weight losses
are also reported for temperature intervals between events in which
usually weight losses occur at a slower rate, and the entire curve
is thus covered by the reported readings.

Authors did not always report readings of the temperatures of
onset and completion (extrapolated), and since these are frequently somewhat
ill-defined they are not reported in the tables. Also, the selection
of temperature ranges along the entire curves is rather subjective,
but it was tried to make the selection in a meaningful way, allowing
a ready comparison of the results obtained from different laboratories.
The choices are not always identical to those made by the authors.

The best way to compare results from different laboratories is
to compare weight losses in those intervals in which a major event takes
place, such as dehydroxylation, loss of CO_2 from carbonates, or loss of
water of crystallization. Losses of adsorbed water, particularly from
the smectites are erratic since storage and pretreatment conditions
of the samples varied widely between the laboratories.

In spite of a generally considerable spread in the data from
different laboratories, in many instances reasonable agreement was
observed when specific features were considered.

In selected cases, a semi-quantitative treatment of the results
could be applied to estimate amounts of specific impurities in the
samples.

Comments on the results for individual samples :

01 Montmorillonite

Loss of structural OH in the range of 500-700°C varies between
3.00 and 3.30% with a mean of 3.20%. Corrected for loss of adsorbed
water below this range, losses of OH vary between 3.40 and 4.10%
with a mean value of 3.75% based on dry weight.(last column of table)
This OH loss is well below the theoretical value of about 5 % , hence
it is likely that some structural OH is already released immediately
below the range in which the major event takes place, together with
residual adsorbed water. This is also supported by the slight slope
of the curve in the intermediate range between desorption of adsorbed

water and release of OH from the structure. A low value for the loss
of structural OH also could indicate the presence of impurities which
do not contribute to weight loss in the same range. Since it is difficult to establish how much of the weight loss is actually due to
loss of structural OH, no firm estimate can be made of the amount of
impurities. Moreover, some impurities may also contribute loss of
structural OH such as other phyllosilicates, and those which yield
a large amount of structural OH such as kaolinite would upset the
calculation considerably.Hence, in most cases, a semi-quantitative
interpretation of TG results can only be done in retrospect when
amount and kind of impurities are estimated from other information.
In the montmorillonite sample such information suggests a total
weight loss of structural OH of roughly 4.50% instead of 5.0%, hence
up to 0.75% of structural OH is probably released before the main
event manifests itself.

02 Laponite

The rather steep slope of the curve between the desorption of
adsorbed water and the release of structural OH (600-800°C range)
suggests that in this intermediate range the loss of structural OH
concurrent with the release of residual adsorbed water is significant. This explains both the smaller than theoretical loss of structural OH in the 600-800°C range and the higher than theoretical loss
in the 200-800°C range (see last column of the table). Impurities
are virtually absent in this synthetic material.

03 Kaolinite (China Clay)

The loss of structural OH amounts to 12.65% (mean value), or
about 90% of the theoretical loss of 14%. Since the principal
impurity appears to be mica, the difference   would
indicate the presence of about 15% mica since this mineral yields
a loss of   about 5% structural OH.

04 Attapulgite

The series of weight losses as reported are generally attributed to
subsequent losses of "free water" (adsorbed on exterior surfaces and
in the channels) up to about 300°C; "bound water", from 300-500/600°C,
and structural OH above 500/600°C. Mean weight losses in these regions
are respectively (7.0 + 2.6) = 9.6% of "free water", 4.2% of "bound
water", and 2.0% of structural OH. However, interpretations of TG
curves in the literature vary, and also losses of the various
categories of water probably overlap. The theoretical total loss

of weight is about 19% (8.5% "free water", 8.5% "bound water", and
2.0% structural OH). The observed mean total weight loss is 15.8%,
which would indicate the presence of at least 17% of other minerals not
contributing to weight loss, i.e. quartz which is present in the sample
to a considerable amount. Since also carbonates and gibbsite are
present, which contribute also to the total weight loss, the amount
of quartz will be even higher than 17%.

05 Illite

Weight losses in the range between 100 and 700°C average 4.0%
corrected for adsorbed water, or to about 80% of the theoretical
value for loss of structural OH, but only 10% quartz and silica
have been estimated to be present in the sample. Perhaps some
additional structural OH is lost both below 100°C and above
600-700°C.

06 Chrysotile

The loss observed in the range between 525 and 725°C is 11.50%.
due to structural OH. In a semi-quantitative analysis based on TG
curves, the weight loss in this region for pure chrysotile is taken
to be 11.77%. Hence the sample would be 97.7% pure chrysotile,
according to one laboratory (GB36). By analogous semi-quantitative
procedures it was found that the sample contains 0.3% brucite,
0.4% magnesite, and 0.25% magnetite (GB36).

07 Crocidolite

Weight losses are minor; those in the higher temperature ranges
are attributed to decomposition of the amphibole with loss of
structural OH.

08 Talc

The total weight loss is primarily due to loss of structural OH.
The somewhat high mean value (5.40% instead of about 5%) can be
explained by the presence of about 10% of a chlorite which gives a
higher structural OH loss than talc.

10 Gibbsite

The observed total weight loss is 34.15 (mean of the two higher value
values) which is close to the theoretical loss of 34.62% for
$Al_2O_3 \cdot 3H_2O$, hence the sample does not contain a significant amount
of impurities.

GB25 carried out a water determination on this sample and
observed a total water loss of 33.8%. Destruction of the trihy-

drate crystal structure with evolution of about 85% of the total
water takes place around 280-300°C. The remaining water is evolved
at 520-550°C when the crystalline boehmite formed at 280-300°C is
finally decomposed. No graph is shown for gibbsite.

11 Magnesite

This sample is rather impure since the observed mean weight loss
of 46.3% is considerably lower than the theoretical weight loss of
52.18%.

12 Calcite

This is a rather pure sample, the observed and theoretical
weight loss being very close (44.2% and 43.97% respectively).

13 Gypsum

The low-temperature weight loss in the formation of the hemihydrate and the anhydride is 19.0% or about 93% of the theoretical
loss of 20.35%, indicating the presence of about 7% impurities,
which may be either calcite or dolomite, and which are responsible
for the additional weight loss around 800°C.

INFRARED SPECTROSCOPY

V. C. Farmer

INTRODUCTION

As the infrared spectrum of a mineral yields information which is
largely complementary to that provided by X-ray and thermal analysis studies,
it must be considered an essential characteristic property of the mineral.
The spectrum can not serve only as a finger-print for identification, but
can also give unique information on features of the structure, including the
family of minerals to which the specimen belongs, the nature of isomorphic
substituents, the distinction of molecular water from constitutional hydroxyl,
the degree of regularity in the structure, and the presence of both crystalline
and non-crystalline impurities. To serve these purposes certain precautions
are necessary in the preparation of the mineral for examination, and in the
capabilities of the spectrometer used to record the spectra. The minimum
conditions for adequate spectra are summarized below; where possible, the
standards defined for Class II reference spectra by the Coblentz Society
(1966) should be aimed at. General guidance on infrared methods is given
by White (1964), and some considerations which particularly affect silicate
spectra are reviewed by Farmer (1964) and Farmer and Russell (1967).

Preparation of the samples

The particle size of the mineral should be sufficiently small to
prevent distortion of the spectra by the Cristiansen effect. Using the KBr

pressed disk technique, a size distribution less than 5μ is often acceptable.
Other techniques (e.g. Nujol mulls) may require particle sizes less than 1μ m.
Clay samples separated by sedimentation are already of satisfactory size, but
require either freeze-drying, or washing with alcohol followed by benzene to
prevent drying out in an intractable form. Minerals which require grinding
should be moistened with water or alcohol (Tuddenham and Lyon, 1960; Farmer,
1964), as dry grinding can destroy the crystal structure.

The alkali halide pressed disk technique is the most generally useful
as a first step in characterizing minerals. More than one concentration of
sample should be used so as to record both weak and strong absorption bands
under optimum conditions. The lowest concentration should be such that
the strongest bands of the spectrum give between 5 and 15 % transmission; a
further spectrum should always be run at five to seven times this concentration. Suitable concentrationsfor silicate spectra correspond to about
0.3 mg and 2 mg sample in 170 mg KBr or 250 mg CsI pressed in 12 mm diameter
disk. Provided the mineral is stable, the prepared disk should be heated at
$100^{\circ}C$, or higher, overnight to reduce the amount of water adsorbed on the
KBr and the sample. Certain samples could require a drying process at a
higher temperature than $100^{\circ}C$; if this is the case, the applied process
should be described.

Samples with cation exchange properties should preferably be saturated
with potassium ions, as polyvalent cations retain coordinated water to high
temperatures.

Other methods of sample preparation e.g. self-supporting films (Fripiat,
Chaussidon and Touillaux, 1960; Farmer and Mortland, 1965; McDonald, 1958;
Angell and Schaffer, 1965) are appropriate for studying surface properties
and thermal reactions of minerals. Some features can best be studied on
single cleavage flakes (e.g. OH bands of certain micas).

Choice of spectrometer and recording of spectra

The fundamental vibrations of minerals can extend from near 4000 cm^{-1}
(OH stretching) down to at least 70 cm^{-1}, and spectra should cover as much

of this range as is possible. The minimum coverage to be aimed at for silicate
minerals is 4000–400 cm^{-1}, with a resolution of 2–3 cm^{-1} throughout the range,
and a precision of $\pm$ 1 cm^{-1} for sharp absorption bands. These standards are
readily achieved with modern grating instruments. Prism instruments give
acceptable spectra below 1600 cm^{-1}, but cannot give sufficient resolution and
precision in the 4000 to 2000 cm^{-1} region. The operator should confirm the
frequency calibration of his instrument from standard data (e.g. those
published by IUPAC, 1961).

It should be remembered that the apparent calibration of a spectrometer
depends on the rate at which the spectrum is scanned. The scanning speed used
should be low enough for the recorder to follow closely the true absorption
curve. The greatest accuracy in measuring the frequency of band maxima,
however, is obtained by operating under static conditions, i.e. by taking
point by point measurements over a sharp absorption band, preferably
symmetrical. Two points having the same height on opposite sides of
the maxima are measured, and the mean of the two wavelengths taken.
This technique should be used both for sharp maxima in the spectra of the
samples studied, and for calibration of the instrument in the region of
these maxima.

Spectra should be plotted on a linear wavenumber (cm^{-1}) scale. A suitable
choice of presentation is as follows:

Abscissa:

2000 cm^{-1} and below	100 cm^{-1} = $2\frac{1}{2}$ cm or 1 inch
4000–2000 cm^{-1}	250 cm^{-1} = $2\frac{1}{2}$ cm or 1 inch
3800–3500 cm^{-1} (for sharp bands)	50 cm^{-1} = $2\frac{1}{2}$ cm or 1 inch

Ordinate:

Linear in transmittance, 0–100 % : 12 cm or 5 inches.

Band maxima should be labelled on the spectra with their frequencies
accurate to 1 cm^{-1} where possible. A line thickness of about 1 mm permits
photographic reduction by a factor of three or more.

References

Angell, G.L. and Schaffer, P.C. (1965), Infrared spectroscopic investigation
    of zeolites and adsorbed molecules. J.Phys.Chem., 69, 3463

Coblentz Society Board of Managers (1966), Specifications for evaluation of
    infrared spectra. Analyt.Chem., 38 (9), 27A

Farmer, V.C. (1964), Infrared spectroscopy of silicates and related compounds.
    In: The Chemistry of Cements, H.P.W. Taylor, Editor. Vol.2, p. 289 London,
    Academic Press

Farmer, V.C. and Mortland, M.M. (1965), Infrared study of complexes of
    ethylamine with ethylammonium and copper ions in montmorillonite.
    J.Phys.Chem., 69, 683

Farmer, V.C. and Russell, J.D. (1967), Infrared absorption spectroscopy in
    clay studies. Clays and Clay Minerals, 15, 121

de Faubert Maunder, M.J. (1971), Practical hints on infrared spectroscopy.
    London, Adam Hilger.

Fripiat, J.J., Chaussidon, J. and Touillaux, R. (1960). Study of dehydration of
    montmorillonite and vermiculite by infrared spectroscopy. J.Phys.Chem.,
    64, 1234

International Union of Pure and Applied Chemistry (1961), Table of wave-
    numbers for the calibration of infrared spectrometers. London, Butterworths.

McDonald, R.S. (1958), Surface functionality of amorphous silica by infrared
    spectroscopy. J.Phys. Chem., 62, 1168

Tuddenham, W.M. and Lyon, R.J.P. (1950), Infrared techniques in the identifi-
    cation and measurement of minerals. Analyt.Chem., 32, 1630

White, R.G. (1964), Handbook of Industrial Infrared Analysis. New York,
    Plenum Press.

CMS RESULTS

Two sets of spectra were submitted:

V.C.Farmer, The Macaulay Institute for Soil Research, Craigiebuckler,
Aberdeen, U.K.

Samples were measured as received, embedded in 0.5 inch thick
KBr disks at one or two different concentrations, and used as such
or after heating at 200°C. Instrument: Perkin Elmer, Model 577,
range 4000-200 cm^{-1}, limited to 250 cm^{-1} by KBr disk.

M.Mortland, Department of Crop and Soil Science, Michigan State
University, East Lansing, Mich.,USA.

Samples were converted to the potassium form and a fraction
< 2 μm was used. The air dried material was embedded in KBr disks
which were dried overnight at 110°C. Instrument: Beckman IR-7,
range 4000-600 cm^{-1}.

Spectral tracings are shown and the observed peak frequencies
are listed in tables, together with assignments (by V.C.Farmer)

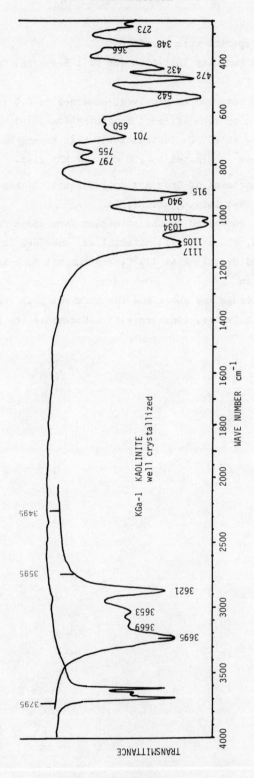

KAOLINITE
KGa-1
well crystallized

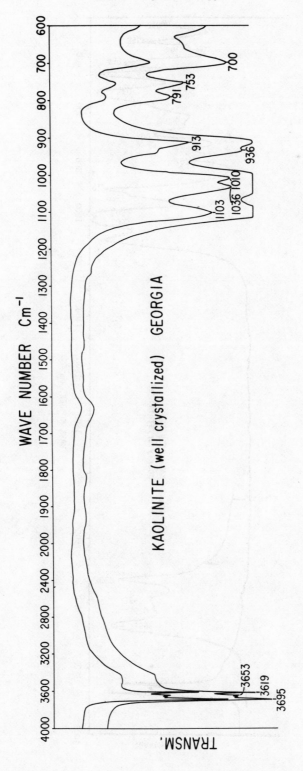

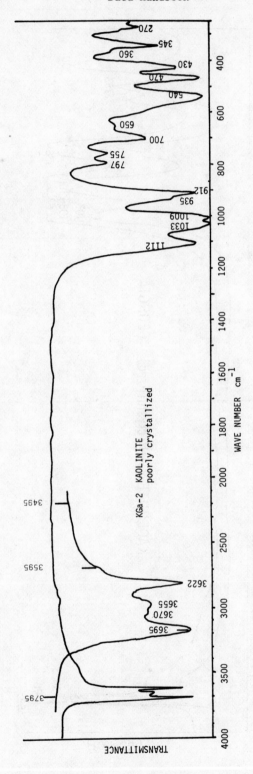

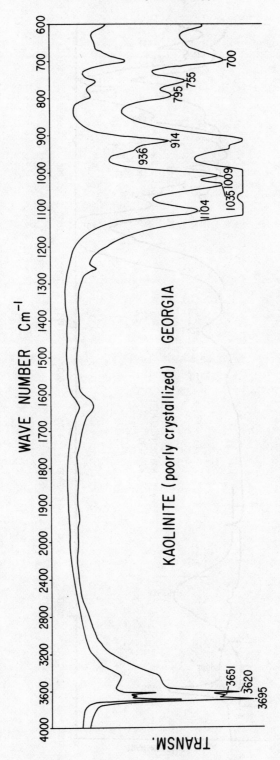

KAOLINITE (poorly crystallized) GEORGIA

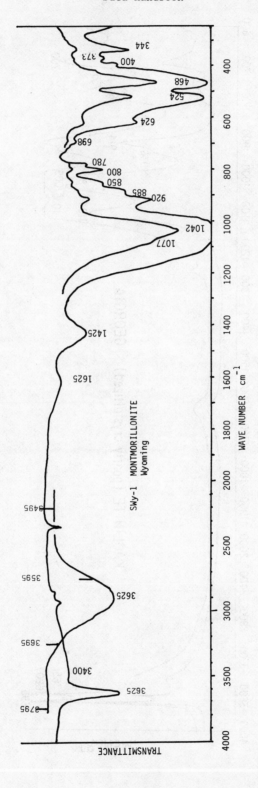

SWy-1 MONTMORILLONITE
Wyoming

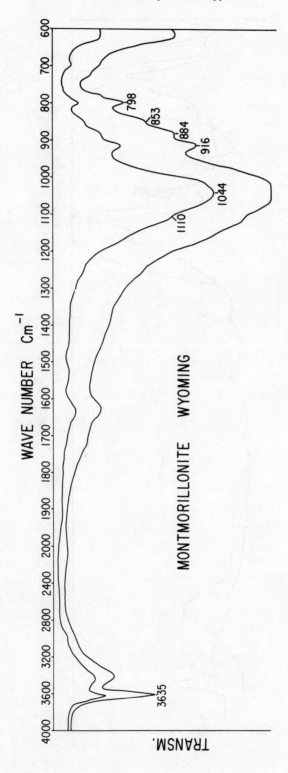

WAVE NUMBER Cm⁻¹

TRANSM.

MONTMORILLONITE WYOMING

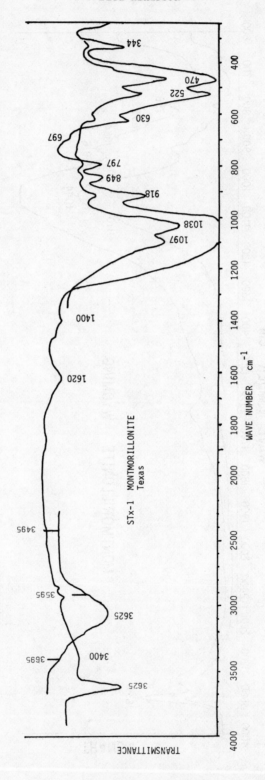

STx-1 MONTMORILLONITE
Texas

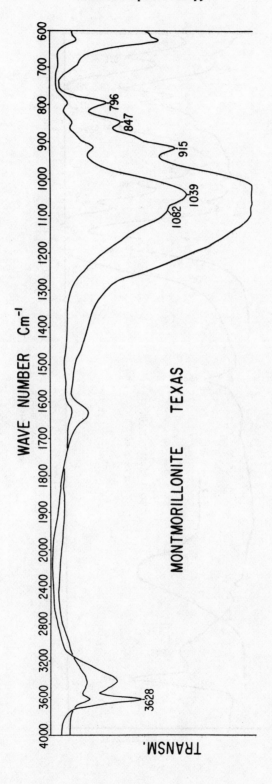

WAVE NUMBER Cm⁻¹

MONTMORILLONITE   TEXAS

TRANSM.

796
847
915
1039
1082
3628

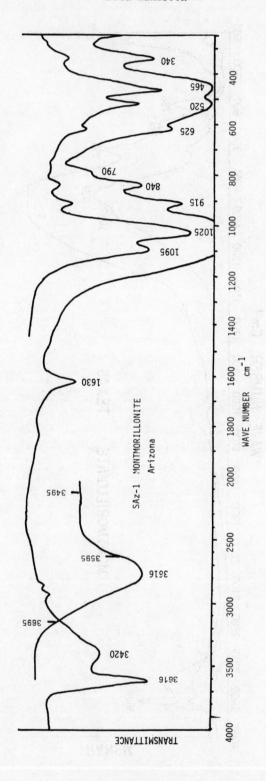

SAz-1 MONTMORILLONITE
Arizona

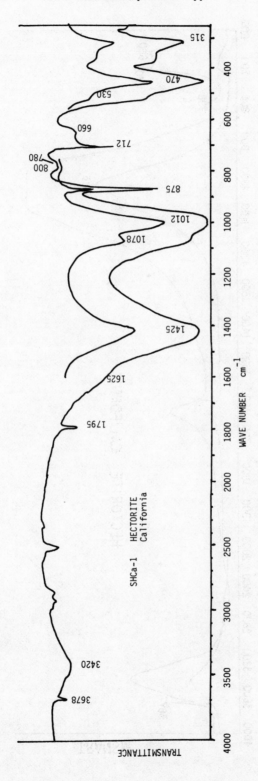

SHCa-1   HECTORITE
California

WAVE NUMBER cm^{-1}

TRANSMITTANCE

Peaks: 315, 470, 530, 660, 712, 780, 800, 875, 1012, 1078, 1425, 1625, 1795, 3420, 3678

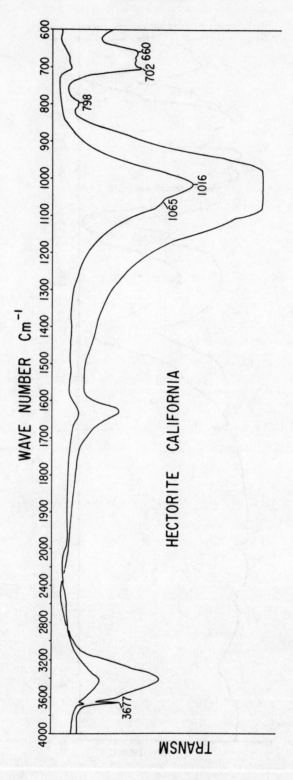

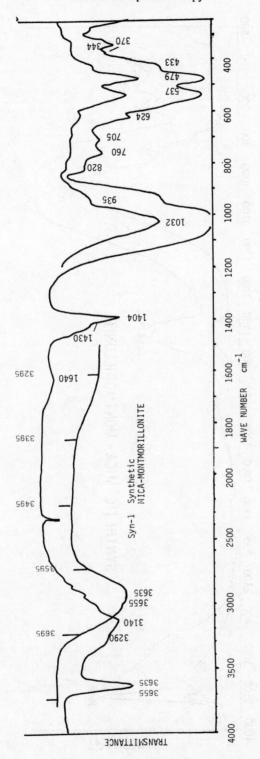

Syn-1 Synthetic
MICA-MONTMORILLONITE

WAVE NUMBER cm⁻¹

TRANSMITTANCE

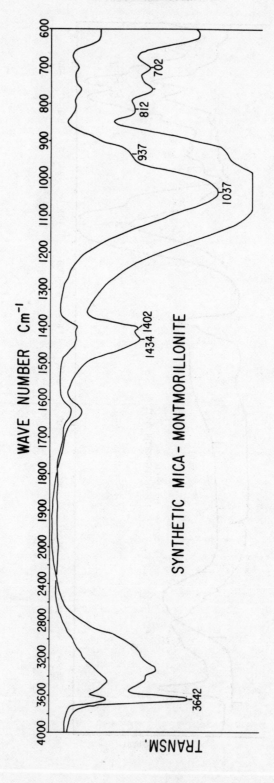

WAVE NUMBER Cm⁻¹

SYNTHETIC MICA-MONTMORILLONITE

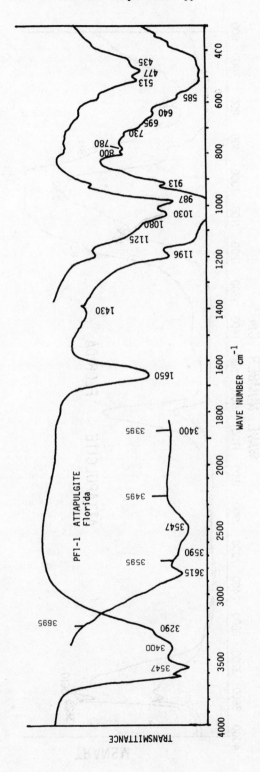

PFI-1 ATTAPULGITE
Florida

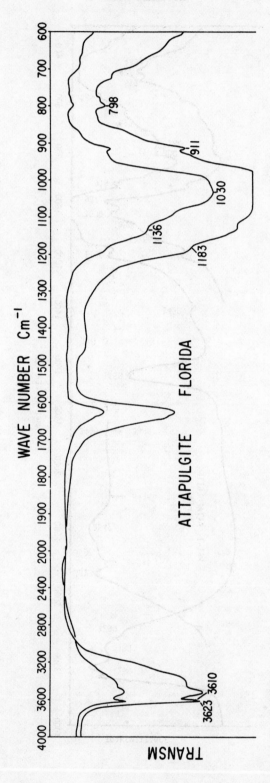

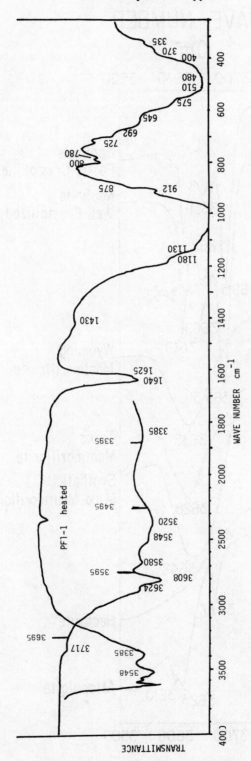

WAVE NUMBER
Cm^{-1}

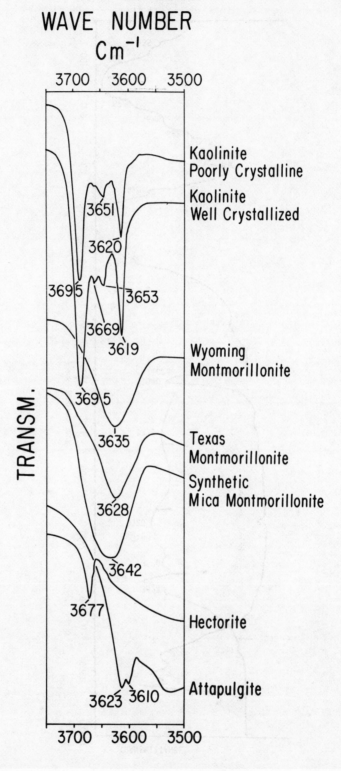

KGa-1 Kaolinite, well crystallized		KGa-2 Kaolinite, poorly crystallized		Assignments
maxima cm^{-1}		maxima cm^{-1}		
K-form <2 μm°	as recvd.°°	K-form <2 μm	as recvd.°°	
3695	3695	3695	3695)	OH stretching,
3669	3669		3670)	hydroxyl sheet
3653	3653	3651	3655)	
3619	3621	3620	3622	OH stretching of inner OH
3410		3400		hydration, OH stretching
1620		1620		hydration, HOH deformation
	1117		1112)	
1103	1105	1104	)	
1036	1034	1035	1033)	SiO stretching
1010	1011	1009	1009)	
936	940	936	935	OH deformation
913	915	914	912	OH deformation
791	797	795	797	
753	755	755	755	
700	701	700	700	
650	650	650	650)	
	542		540)	mixed SiO deformations
	472		470)	and octahedral sheet
	432		430)	vibrations
	366		360)	
	348		345)	
	273		270)	

(°) 0.21 and 1.06 mg/cm^2, dried at 110°C in KBr disk

(°°) 0.5 mg, heated at 200°C in 0.5 inch KBr disk

SWy-1   Montmorillonite, Wyoming

maxima   cm^{-1}

K-form <2 μm°	as recvd.°°	Assignments
3635	3625	OH stretching
3400	3400	hydration OH stretching
1625	1625	hydration, HOH deformation
	1425	carbonate
1110	1077)	SiO stretching
1044	1042)	
916	920	OH deformation, linked to 2 Al^{3+}
884	885	OH deformation, linked to Fe^{3+}, Al^{3+}
853	850	OH deformation, linked to Al^{3+}, Mg^{2+}
798	800)	
	780)	silica, quartz
	698)	
	624	
	524)	SiO deformation and
	468)	AlO stretching
	400)	
	373)	quartz
	344	

(°) 0.21 and 1.06 mg/cm^2 dried at 110°C in KBr disk

(°°) 0.33 mg in 0.5 inch KBr disk and 2 mg heated at 200°C in
0.5 inch KBr disk

STx-1   Montmorillonite, Texas

maxima   cm^{-1}

K-form <2 μm°	as recvd.°°	Assignments
3628	3625	OH stretching
3400	3400	hydration, OH stretching
1620	1620	hydration, HOH deformation
	1400	carbonate
1082	1097)	SiO stretching
1039	1038)	
915	918	OH deformation, linked to 2 Al^{3+}
847	849	OH deformation, linked to Al^{3+} Mg^{2+}
796	797	silica
	697	quartz
	630)	
	522)	SiO deformation and
	470)	AlO stretching
	344)	

(°) 0.21 and 1.06 mg/cm^2 dried at 110°C in KBr disk

(°°) 0.33 mg in 0.5 inch KBr disk and 2 mg heated at 200°C in
0.5 inch KBr disk

SAz-1 Montmorillonite, Arizona

maxima cm^{-1}

as recvd.[oo]	Assignments
3616	OH stretching
3420	hydration, OH stretching
1630	hydration, HOH deformation
1095) 1025)	SiO stretching
915	OH deformation, linked to 2 Al^{3+}
840	OH deformation, linked to $Al^{3+}Mg^{2+}$
790	silica
625) 520) 465) 340)	SiO deformation and AlO stretching

([oo]) 0.25 and 2.0 mg heated at 150°C in 0.5 inch KBr disk

SHCa-1   Hectorite, California

maxima cm^{-1}

K-form <2 μm[o]	as recvd.[oo]	Assignments
3677	3678	OH stretching
3400	3420	hydration, OH stretching
	1795	calcite
1625	1625	hydration, HOH deformation
	1425	calcite
1065	1078)	SiO stretching
1016	1012)	
	875	calcite
798	800)	silica
	780)	
	712	calcite
702		quartz
660	660	OH deformation
	530	MgO stretching out of plane
	470	SiO deformation in the plane
	315	calcite

([o]) 0.21 and 1.06 mg/cm^2 dried at 110°C in KBr disk

([oo]) 0.33 mg in 0.5 inch KBr disk, and 2 mg heated at 200°C in 0.5 inch KBr disk

Syn-1 Synthetic mica-montmorillonite

maxima cm^{-1}

K-form <2 μm°	as recvd.°°	Assignments
	3655	OH stretching
3642	3635	OH stretching
3400		hydration, OH stretching
3300	3290)	NH_4^+
3150	3140)	
1630	1640	hydration, HOH deformation
1434	1430)	NH_4^+
1402	1404)	
1037	1032	SiO stretching
937	935	OH deformation, linked to 2 Al^{3+}
812	820	
760	760	
702	705	quartz
	624)	
	537)	
	479)	mixed SiO deformation and
	433)	AlO stretching
	370)	
	344)	

(°) 0.21 and 1.06 mg/cm^2 dried at 110°C in KBr disk

(°°) 0.33 mg in 0.5 inch KBr disk and 2 mg heated at 200°C in
0.5 inch KBr disk

PF1-1   Attapulgite, Florida

maxima cm^{-1}

K-form <2 μm°	as recvd.°°	heated°°°	Assignments
		3717)	
3623		3624)	
3610	3615	3608)	OH stretching
3580	3590	3580)	
3535	3547	3548)	
		3520)	
3400	3400	3385)	hydration, OH stretching
3290	3290	3215)	
1650	1650	1640)	hydration, HOH deformation
		1625)	
1430	1430	1430	calcite
1183	1196	1180)	
1136	1125	1130)	
1090	1080	N.D)	SiO stretching
1030	1030	N.D)	
	987	N.D)	
911	913	912	OH deformation, linked to Al^{3+}
870	870	875	calcite
798	800	800)	
775	780	780)	quartz
	730	725	
	695	692	quartz
	640	645)	mixed tetrahedral deformation
	585	575)	and octahedral vibrations
	513	510)	
	477	480)	
	435	400)	
		370)	
		335)	

(°) 0.21 and 1.06 mg/cm^2 dried at 110°C in KBr disk

(°°) 0.33 and 2 mg in 0.5 inch KBr disk

(°°°) 2 mg heated at 200°C in 0.5 inch KBr disk

CMS COMMENTS (V. C. Farmer)

KGa-1 and KGa-2, Kaolinite

The two kaolinites are not very extreme samples of the range of order that exists in kaolinites. The "well crystallized" sample is not as good as typical China clay from Cornwall as judged by the intensity of the 3669 cm^{-1} band which is peculiar to well-crystallized kaolinites. The "poorly crystallized" sample still shows this band, which is absent from more disordered kaolinites, such as the type specimen from Pugu.

The "well crystallized" sample shows a splitting of the 1100 cm^{-1} band, a feature commonly found for fairly coarse crystals. This splitting is not shown in the small particle size fraction. The splitting is due to particle size and shape and its cause is well understood. (see Farmer,V.C. and Russell,J.D., Spectrochimica Acta, 22, 389-398, 1966).

SWy-1 Montmorillonite, Wyoming

From its infrared spectrum this is a fairly typical Wyoming bentonite with the usual moderate Fe^{3+} content (band at 885 cm^{-1}). Quartz impurity bands are present at 780, 800, 698, 400, and 373 cm^{-1}. The fairly strong band at 800 cm^{-1} indicates the presence of disordered tridymite (see also comments on the STx-1 spectrum). A trace of carbonate (band at 1425 cm^{-1}) is apparently eliminated in the fraction <2 μm.

STx-1 Montmorillonite, Texas

Judging from the infrared spectrum this bentonite has a low iron content, as there is little absorption at 880 cm^{-1} due to (Fe Al OH) groups.

A band at 797 cm^{-1} is unusually strong for a montmorillonite. It is probably due to a silica phase. A similar band shown by an Italian montmorillonite was observed by Russell and recognized to be due to a clay-size platy disordered tridymite· (see: Wilson,M.J., Russell,J.D., and Tait, J.M.,A new interpretation of the structure of disordered α-cristobalite. Contrib. Mineral. Petrol., 47, 1-6, 1974.)

SAz-1 Montmorillonite, Arizona

From its spectrum, this appears to be a clean montmorillonite with a very low octahedral iron content, similar to the Cheto-type montmoril-

lonite examined by Grim and Kulbicki (Am.Miner., $\underline{46}$, 1329–1369, 1961),or
the Skyrvedalen montmorillonite examined by Farmer and Russell (Spectrochimica Acta, 20, 1149–1174, 1964). It is distinguished from these by the
presence of a weak band at 790 cm^{-1} indicative of traces of silica (see
above)

SHCa-1 Hectorite, California

Absorption bands of calcite appear strongly in the spectrum of
the crude sample. There may also be a little quartz present,
contributing weak features at 800 and 780 cm^{-1}, superimposed
on a weak band of hectorite. In the <2 µm fraction calcite bands
are completely absent and an apparently pure hectorite is obtained.

Syn-1 Synthetic mica-montmorillonite

The infrared spectrum is broadly similar to that of muscovite,
but more diffuse. NH_4 absorption bands appear at 1432 and at 1404
cm^{-1}. Much of the NH_4 has migrated from the clay into the KBr,
resulting in a spectrum of NH_4Br in solid solution in KBr (bands
at 1404 and 3100 cm^{-1}). The only feature in the spectrum which
indicates a relationship with montmorillonite (apart from exchangeable NH_4^+) is the band at 624 cm^{-1}. The spectrum is nearer
to that of muscovite than that of beidellite.

PF1-1 Attapulgite

The infrared spectrum indicates a fairly pure attapulgite with
possibly a trace of quartz giving features at 780 and 800 cm^{-1}.
The absorption bands are not as sharp as those given by the
best specimens, for example that in the OECD collection.

The spectrum shows considerable modification on drying, even
at 110°C: band positions shift in the OH stretching region (3000–
3700 cm^{-1}) and the Si-O stretch (1000–1200 cm^{-1}). All these shifts
are reversible and indicate a distortion of the silicate framework
as zeolite water is removed. Further changes are known to occur
at higher temperatures as coordinated water is removed.

OECD RESULTS (including Raman)

Spectra, or tabulated results were received for all samples. The
number of laboratories submitting results for each sample varied
from one to five. For most samples three sets of results were obtained.
In all,nine laboratories participated. One laboratory (B1) gave particularly

complete high quality spectra covering the range of 4000–400 cm^{-1} for
all samples. These spectra are reproduced in the figures. For all
samples observed frequencies of absorption bands are listed in the
tables, together with assignments. Ranges of frequencies observed are
given whenever variations larger than 5 cm^{-1} were reported.

Raman spectra were submitted by one laboratory for magnesite,
calcite, and gypsum. Raman frequencies are included in the relevant
IR frequency tables.

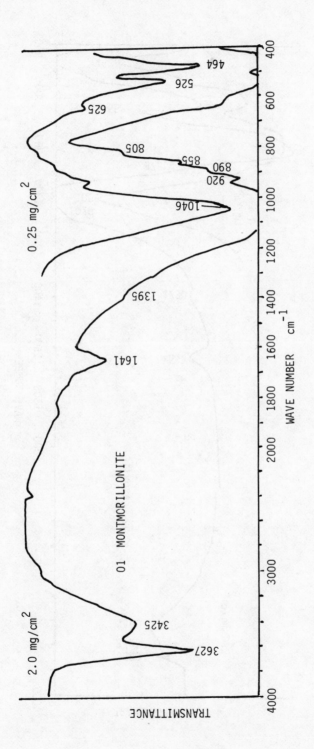

01 MONTMORILLONITE

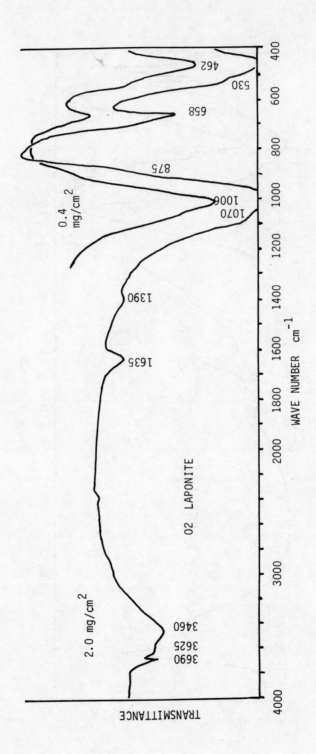

02 LAPONITE

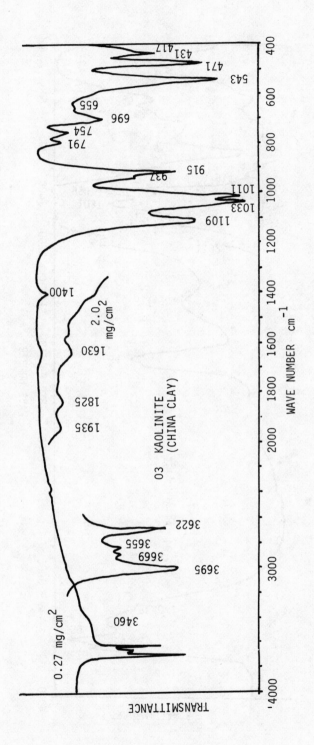

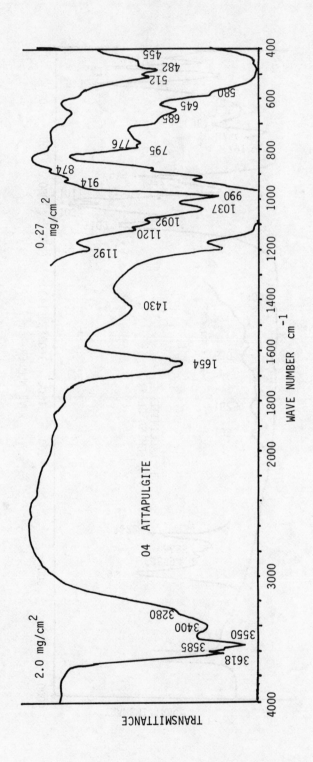

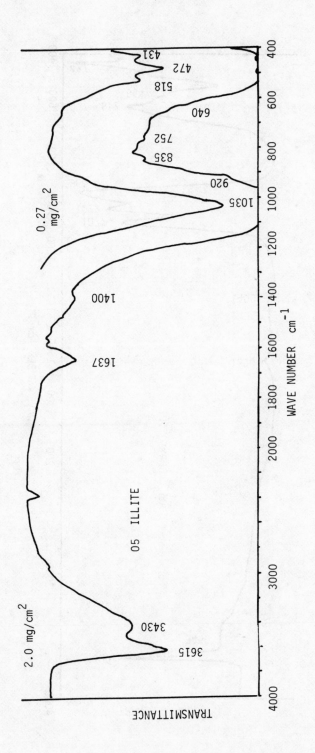

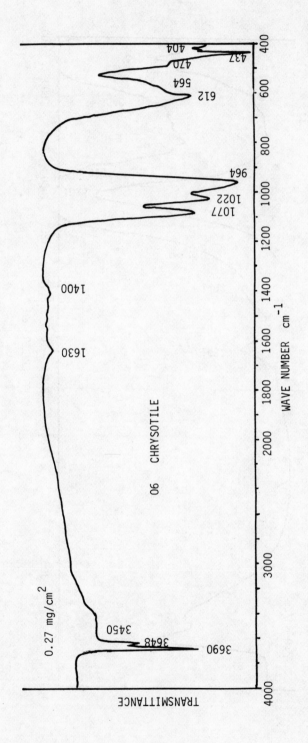

06   CHRYSOTILE

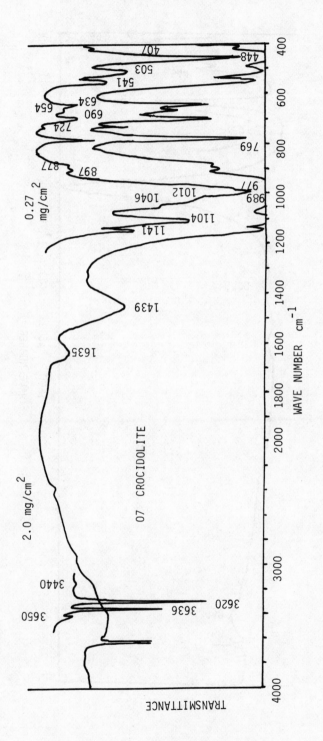

07 CROCIDOLITE

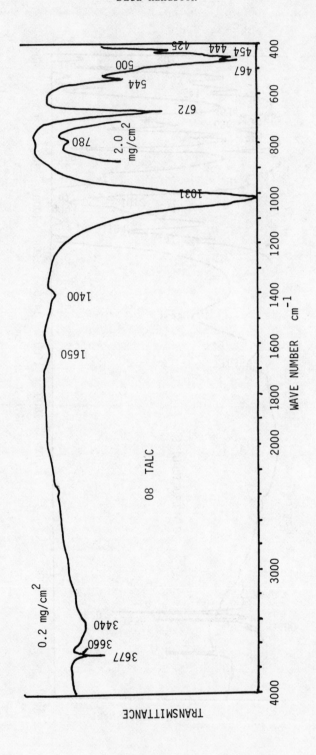

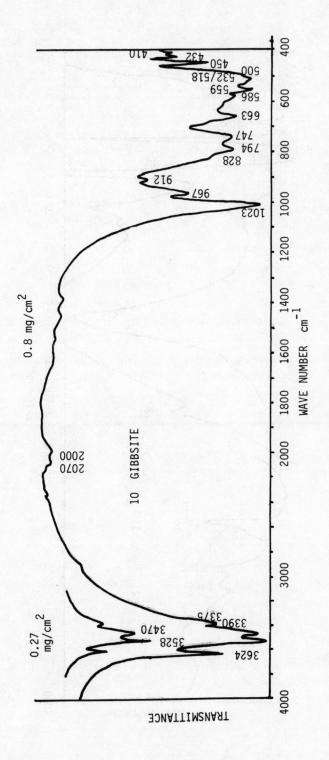

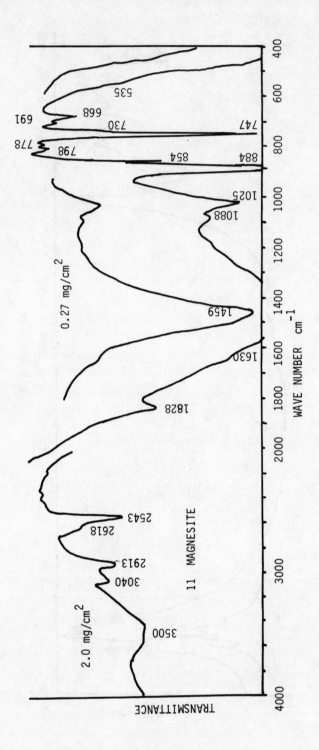

11 MAGNESITE

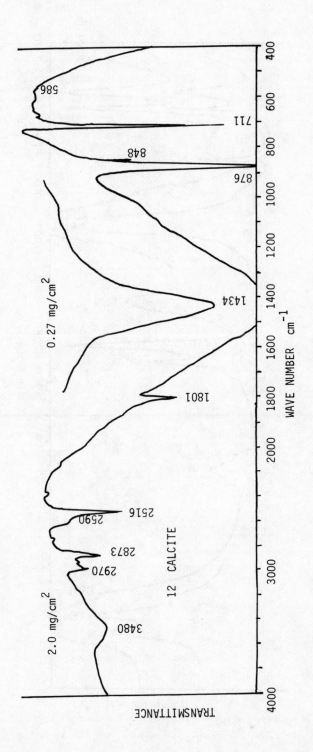

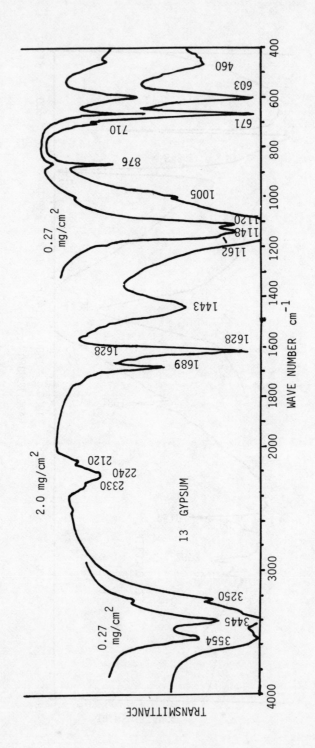

13 GYPSUM

OECD IR

maxima cm^{-1}	assignments	maxima cm^{-1}	assignments
01 Montmorillonite			
3628	OH stretching	918	OH deformation, linked to $2Al^{3+}$
3425-3440	hydration, OH stretching	890	OH deformation, linked to Fe^{3+}, Al^{3+}
1635	hydration, HOH deformation	845-855	OH deformation, linked to Al^{3+}, Mg^{2+}
1395	carbonate (trace)	795-805	
1112	SiO stretching in the plane	695	
1086	SiO stretching out-of-plane	627	
1037-1045	SiO stretching	520-530)	
		468)	SiO deformation
		438)	

maxima cm^{-1}	assignments	maxima cm^{-1}	assignments
02 Laponite			
3690)		875	
3664?)	OH stretching	700	SiO deformation out-of-plane
3625)		654-665	OH deformation
3455	hydration, OH stretching	537	MgO deformation out-of-plane
1635	hydration, HOH deformation	462	SiO deformation in the plane
1400	carbonate (trace)		and OH translation
1071	SiO stretching out-of-plane		
1005	SiO stretching		

maxima cm^{-1}	assignments	maxima cm^{-1}	assignments
03 Kaolinite (China Clay)			
3692-3700)		937)	OH deformation
3668)		913)	
3655)	OH stretching	790	
3618-3625)		756	
3460	hydration, OH stretching	697	
1638	hydration, HOH deformation	650	
1400	carbonate (trace)	535-543)	
1110	SiO stretching out-of-plane	465-475)	SiO deformation
1033	SiO stretching	433)	
1009	SiO stretching	416	muscovite?

maxima cm^{-1}	assignments	maxima cm^{-1}	assignments
04 Attapulgite			
3617)	structural OH and adsorbed	912	OH deformation, linked to Al^{3+}
3585)	water OH stretching; upon	875	calcite
3550)	drying at 500°C replaced by	795	quartz
3395)	narrower bands at 3625,3610,	776	quartz
3280)	3580,3515,3385,3214	692	quartz
1652	hydration, OH deformation; on	642	
	drying shift to 1630	580	
1430	carbonate (trace)	510	
1195)		480	
1122)		455	
1091)	SiO stretching; become		
1035)	diffuse upon drying		
987)			

maxima cm^{-1}	assignments	maxima cm^{-1}	assignments

05 Illite

maxima cm^{-1}	assignments	maxima cm^{-1}	assignments
3618	OH stretching, linked to $2Al^{3+}$	920	OH deformation, linked to $2Al^{3+}$
3580	OH stretching, linked to Al^{3+}, Fe^{3+}	880	OH deformation, linked to Al^{3+}, Fe^{3}
3400-3490	hydration, OH stretching	835	OH deformation, linked to Al^{3+}, Mg^{2}
1637	hydration, HOH deformation	790	impurity
1400	carbonate (trace)	752	
1080	SiO stretching	723	impurity
1025-1035	SiO stretching	640	impurity
		520	
		472	
		430	

06 Chrysotile

maxima cm^{-1}	assignments	maxima cm^{-1}	assignments
3690	OH stretching	612	OH deformation
3650	OH stretching	564	
3450	hydration, OH stretching	470	
1630	hydration, HOH deformation	437	
1400	carbonate (trace)	404	
1077	SiO stretching		
1022	SiO stretching		
964	SiO stretching		

07 Crocidolite

maxima cm^{-1}	assignments	maxima cm^{-1}	assignments
3650	OH stretching, linked to $2Mg^{2+}, Fe^{2+}$	893	
3636	OH stretching, linked to $Mg^{2+}, 2Fe^{2+}$	876	
3620	OH stretching, linked to $3Fe^{2+}$	770	
1635	hydration, HOH deformation	720	
1440	carbonate	690	
1142		655	
1104		635	
1046		541	
1012		503	
989		448	
977		408	

08 Talc

maxima cm^{-1}	assignments	maxima cm^{-1}	assignments
3677	OH stretching, linked to $3Mg^{2+}$	780	
3660	OH stretching, linked to $Fe^{2+}, 2Mg^{2+}$	(690)	SiO deformation out-of-plane
3440	hydration, OH stretching	673	OH deformation
1925		535	MgO stretching out-of-plane
1820		-544	
1650	hydration, HOH deformation	500	SiO deformation in the plane
1400	carbonate (trace)	466	OH translation in the plane
(1045)	SiO stretching out-of-plane	452	SiO deformation
1018	SiO stretching in the plane	442	SiO deformation
		425	SiO deformation

maxima cm^{-1} assignments   maxima cm^{-1} assignments

10 Gibbsite

3620	)	
3524	)	
3460	) OH stretching	
3394	)	
3375	)	
2070		
2000		
1022	)	
969	)	
936	) OH deformation	
915	)	
832		

796
745
664
587
559
(532)
517
502
448
423
410

11 Magnesite

3500 hydration, OH stretching
3020-3040
2900-2913 $2\nu_3$ CO_3^{2-}
2618
2532-2543 $\nu_1 + \nu_3$ CO_3^{2-}
1818-1830 $\nu_1 + \nu_4$ CO_3^{2-}
1430-1460 ν_3 CO_3^{2-} ν_1 CO_3^{2-} (Raman : 1439)
         ν_1 CO_3^{2-} (Raman : 1095)
1088 quartz
1015-1025 talc

884 ν_2 CO_3^{2-}
854   ν_2 CO_3^{2-} for $C^{13}O_3^{2-}$
797 quartz
780 quartz
747 ν_4 CO_3^{2-} (also Raman)
730 dolomite
691-699 quartz
668 talc
535 impurity

12 Calcite

3480 OH stretching, hydration
2970-2980
2872(2841)
2585
2505-2516 $\nu_1 + \nu_3$ CO_3^{2-}
1792-1801 $\nu_1 + \nu_4$ CO_3^{2-}
1420-1434 ν_3 CO_3^{2-} (Raman: 1439)
1088 Raman: $\nu_1 CO_3^{2-}$

876
848
709 ν_4 CO_3^{2-} (Raman: 713 cm^{-1})
(690)
586

(1000-1100) silicate

13 Gypsum

3550	)	
3490	)	
3404	) OH stretching, water of	
3244	) crystallization	
2330		
2240		
2120		
1680-1689	) HOH deformation,water of	
1615-1628	) crystallization	
1440	calcite	

1162)
1145) ν_3 for SO_4^{2-}
1117)
1035 silicate
1005 ν_1 for SO_4^{2-} (Raman: 1010)
980 silicate
876 calcite
710 calcite
669)
) ν_4 for SO_4^{2-}
602)
460 libration of water of
      crystallization

OECD COMMENTS

V.C.Farmer, The Macaulay Institute for Soil Research, Craigiebuckler, U.K.

Generally, the results of different laboratories agreed within the usual variation expected for infrared spectra, or showed differences explicable in terms of pretreatment, or sample preparation, but for one sample at least (gypsum) there is evidence that the distribution of minor contaminants is not uniform. Some of the factors which caused variation in the spectra are listed below.

Mounting technique.

One laboratory dispersed their samples in mineral oil; most used potassium bromide pressed disks. Mineral oil mulls tend to orient platy particles, so that vibrations whose dipole change is perpendicular to the plates absorb more weakly in mulls than in pressed disks. This effect was particularly evident for kaolinite and montmorillonite. The absorption band of mineral oil near 1400 cm^{-1} obscured a band of carbonate, which was a common contaminent of these samples.

Grinding samples.

Some grinding was essential to obtain good quality spectra of most samples, as their particle size was too large. Grinding is often incorporated in the process of mixing the samples with KBr in the preparation of pressed disks, and this was sometimes sufficient to give adequate spectra. With heterogeneous samples, a gentle grind may reduce one component to the optimum size, while leaving the particles of a harder component too large to absorb infrared radiation efficiently. This may account for some variation in the intensity of absorption by a quartz contaminant in the magnesite.

Size fractionation of samples.

A preliminary separation of the finer fractions of a sample may result in the elimination of a contaminant whose particles are all large, or the concentration of a contaminant whose particles are all small. This effect was evident with palygorskite, from which one laboratory separated a fine fraction for examination. This fine fraction was almost free of quartz, which was present in the whole sample, but contained more calcite, which may be an artifact of the sodium carbonate solution used to disperse the sample. A fine fraction from the gibbsite gave sharper spectra than did finely ground samples of the bulk material; this effect may arise from differences in the degree of crystallinity between different size fractions.

Thermal history of the sample.

Gypsum dehydrates to the soluble anhydride at $100^{\circ}C$, which then re-
hydrates in the air to form the hemihydrate. Thus a sample which had been
dried at $120^{\circ}C$ gave the spectrum of the hemihydrate. More unexpected was
the sensitivity of the palygorskite spectrum to its state of hydration,
which affects not only the absorption bands of hydroxyl and water, but
also Si-O absorption bands. The effect is reversible, but the recommended
procedure of recording the spectrum in KBr disks after drying at $100^{\circ}C$
gives the spectrum of a partially dehydrated sample.

Separation of components.

Contaminants present in a sample will usually differ in density,
particle size, and shape. Unless the whole sample is extremely fine,
some separation of the different components is likely during handling,
and the very small sample (about 1 mg) used for infrared spectroscopy
could well be unrepresentative. Differences between the amounts of
calcite and silicate present in the gypsum were indicated by the spectra
of the three laboratories, and this was confirmed by Farmer, who examined
samples supplied by each, using identical sample preparation and
recording conditions. Some of the differences between laboratories
could also be ascribed to the grinding conditions used. The calcite
appeared to be of smaller particle size, and the intensities of its
absorption bands were less enhanced by grinding than were those of
gypsum.

Many of the spectra show a weak doublet at 2860 and 2930 cm^{-1} due
to CH stretching, and a sharp band near 2340 cm^{-1}, due to CO_2,
liberated on heating the disk.

01 Montmorillonite

The spectrum is similar to that of Wyoming bentonite, and other
montmorillonites containing a proportion of octahedral Fe^{3+}, in
that it shows an absorption shoulder at 890 cm^{-1}, due to a bending
vibration of OH coordinated to $AlFe^{3+}$ pairs, in addition to the
bending vibrations of Al_2OH groupings (at 920 cm^{-1}) and AlMgOH
groupings (at 850 cm^{-1}), which are given by all montmorillonites.
Other features of the spectra common to all dioctahedral montmorillonites include OH stretching (3630 cm^{-1}), SiO stretching (1000-
1120 cm^{-1}), SiO bending (430-530 cm^{-1}), and water bands at 3430 cm^{-1}
and 1635 cm^{-1}. The only detectable impurity was a trace of carbonate,

giving a weak band at 1400 cm^{-1}, whose intensity varied with the pretreatment and technique of sample preparation. No kaolinite, for which infrared
spectroscopy is particularly sensitive, could be detected.

The five laboratories submitting results were in good agreement. The
most variable feature in the spectra was the intensity of the SiO stretching vibration at 1086 cm^{-1}, which has a dipole oscillation perpendicular
to the layers. This shoulder was absent from a mull spectrum, because of
the orientation of the sample by this technique. Its intensity varied in
spectra obtained using KBr disks. Because of the absence of the 890 cm^{-1}
band in the mull spectrum, the weak in-plane SiO vibration at 1115 cm^{-1}
was most easily seen there. Absorption bands and shoulders which could
be detected in most spectra are listed in the table. The range of
frequencies found is given where this exceeds 5 cm^{-1}. Discrepancies in
the reported positions can generally be ascribed to the breadth of the
absorption bands.

02 Laponite

The spectrum is close to that of natural hectorite. Features of the
spectrum include OH stretching (3690 cm^{-1}), SiO stretching (1000–1100
cm^{-1}), SiO out-of-plane bending (700 cm^{-1}), OH bending (658 cm^{-1}), an
MgO out-of-plane vibration (537 cm^{-1}), SiO in-plane bending and OH
translation (462 cm^{-1}), and water absorption. The only detectable impurity
was a trace of carbonate, giving a weak band near 1400 cm^{-1}.

Four laboratories submitted results, but only two obtained spectra
of acceptable quality. These showed only minor differences, affecting
principally the intensity of bands due to perpendicular vibrations at
1070 cm^{-1} and 700 cm^{-1}.

03 Kaolinite (China Clay)

The spectrum is that of a well ordered kaolinite. Features of the
spectrum include OH stretching (3620–3700 cm^{-1}), SiO stretching
(1000–1120 cm^{-1}), OH bending (910–940 cm^{-1}), and SiO bending (400–550
cm^{-1}). The only detectable impurities were carbonate (weak absorption
near 1400 cm^{-1}), and possibly muscovite (a weak band at 415 cm^{-1},
reported by one laboratory).

Five laboratories submitted results, all in fair agreement. The
orientation of the sample in a mull led to marked weakening of absorption due to perpendicular vibrations at the following frequencies:
3697, 1110, 756, and 700 cm^{-1}.

04 Attapulgite

This sample gives spectra superior to any published in the literature, in sharpness and definition of its absorption bands, indicating that it has a well-ordered structure. The spectra shown were obtained without heating the disk. Partial dehydration of the sample by drying the disk at 100 $^{\circ}$C causes profound, but reversible changes in the spectrum, affecting both the OH stretching region and the SiO stretching region. The absorption bands of OH stretching (structural OH and H_2O) of the air-dry specimen as listed in the table, are replaced by sharper bands at 3625, 3610, 3580, 3515, 3385, and 3214 cm^{-1}, and H_2O bending shifts from 1650 to 1630 cm^{-1}, as previously noted by Ovcharenko (1966). In the SiO stretching region, most of the sharper absorption bands in the 900-1200 cm^{-1} region weaken and diffuse after drying at 100 $^{\circ}$C; a new band appears at 1127-1147 cm^{-1} and the 1037 cm^{-1} band shifts to 1024 cm^{-1}.

Impurities present include calcite, absorbing at 1430 cm^{-1} and 874 cm^{-1}, and quartz, absorbing at 795, 775, and 690 cm^{-1}. The amount of quartz was greatly reduced in a fine fraction (< 10 μm).

Results of the four laboratories submitting spectra were in good agreement, once the effect of dehydration was recognized. A mull spectrum gave results similar to those obtained by the KBr technique.

05 Illite

The spectrum is typical of a highly disordered ferruginous illite. The stretching and bending vibrations of OH coordinated to two octahedral Al^{3+} gives bands at 3620 cm^{-1} and 920 cm^{-1}. Shoulders on the low-frequency side of these bands, at 3580 cm^{-1} and 880 cm^{-1}, can be ascribed to hydroxyl coordinated to a $AlFe^{3+}$ octahedral pair. This is an unusual feature of illite spectra. The weak band at 835 cm^{-1} is a common feature of the illites and phengites, and probably arises from OH bending of an MgAlOH grouping. The pattern of absorption in the SiO stretching region (1000-1100 cm^{-1}) and bending region (400-550 cm^{-1}) are similar to, but more diffuse than those of muscovite.

No quartz or kaolinite were detectable in the spectra, but weak bands at 640 cm^{-1}, 723 cm^{-1}, and 790 cm^{-1} suggest the presence of an impurity, as these bands are absent from the spectra of other illites. A trace of carbonate, absorbing at 1400 cm^{-1}, is present

Results from the four laboratories submitting spectra were in good agreement.

06 Chrysotile

A typical chrysotile spectrum. The reason for the doublet in the OH stretching region, and for the complexity of the SiO stretching region is not well understood. Bending vibrations of OH probably account for most of the strong absorption in the 600 cm^{-1} band, by analogy with talc.

The only detectable impurity is a trace of carbonate, absorbing at 1400 cm^{-1}.

Results from two laboratories were in good agreement. A third laboratory had difficulty in sample preparation.

07 Crocidolite

Only one laboratory submitted a spectrum, which is entirely typical of crocidolite, apart from a little carbonate absorbing at 1440 cm^{-1}. The three OH stretching bands arise from OH coordinated with respectively Fe_3^{2+} (3620 cm^{-1}), $Fe_2^{2+}Mg$ (3636 cm^{-1}), and $Fe^{2+}Mg_2$ (3650 cm^{-1})

08 Talc

The spectrum is in excellent agreement with published spectra; there is no detectable impurity apart from adsorbed water. A small octahedral Fe^{2+} content is indicated by the weak OH stretching band at 3661 cm^{-1}, due to $Fe^{2+}Mg_2$ groupings. The OH bending vibration lies at 672 cm^{-1}. Vibrations of the silicon tetrahedral sheet perpendicular to the layers at 1045 and 690 cm^{-1} are poorly resolved in these spectra. In-plane SiO vibrations occur at 1020 cm^{-1} and in the 400–500 cm^{-1} region. A perpendicular vibration involving Mg lies at 545 cm^{-1}, and an in-plane translatory vibration of OH at 466 cm^{-1}.

Only two laboratories submitted spectra which were in good agreement.

10 Gibbsite

The spectrum indicates a pure gibbsite. Pre-ground material and a fine fraction separated by sedimentation gave sharper spectra than the whole sample which does not appear to be so well crystallized as the best natural samples. From a comparison with synthetic $Al(OD)_3$, the bands at 900–1050 cm^{-1} are known to be bending vibrations. No impurities were present.

Results from the three laboratories were in good agreement.

11 Magnesite

The spectrum indicated the presence of a number of impurities. In the whole sample, bands of talc (1025 and 668 cm^{-1}), quartz (1088, 798,

778, and 695 cm^{-1}), and dolomite (732 cm^{-1}) can be seen. Some black grains isolated from the whole sample proved to contain muscovite and quartz.

One laboratory (B3) also obtained a Raman spectrum, but found only a single line, at 1095 cm^{-1}. This laboratory gave a very complete assignment for the absorption of the carbonate ion in magnesite: ν_1: 1095 cm^{-1}; ν_2: 880 cm^{-1}; ν_3: 1439 cm^{-1}; ν_4: 745 cm^{-1}; $\nu_1+\nu_4$: 1818 cm^{-1}; $\nu_1+\nu_3$: 2532 cm^{-1}; $2\nu_3$: 2899 cm^{-1}. The ν_2 vibration of the $^{13}CO_3$ ion in natural abundance is a very sharp band at 854 cm^{-1}.

Results from the three laboratories submitting spectra were in fair agreement.

12 Calcite

The spectrum indicates an almost pure calcite; weak absorption in the 1000–1100 cm^{-1} region may indicate a very small silicate contaminant.

Three laboratories provided spectra in good agreement, and one of these also obtained a Raman spectrum, showing bands at 713 cm^{-1} (ν_4), 1088 cm^{-1} (ν_1), and 1439 cm^{-1} (ν_3). The assignments of the bands are analogous to those for magnesite.

13 Gypsum

A typical gypsum spectrum, with calcite contaminant, absorbing at 1445, 876, and 710 cm^{-1}, and a silicate contaminant giving broad absorption near 980 and 1035 cm^{-1}. The intensity of the contaminant bands varied in the three spectra submitted. This variation can partly be ascribed to differences in the degree of grinding during sample preparation, but seems also to be due to a real variation in the proportion of contaminants in the three different samples.

A Raman spectrum which was difficult to obtain, showed only the symmetric stretching (ν_1) of the sulphate ion at 1010 cm^{-1}. Other assignments for the sulphate ion include the three components of ν_3 at 1164, 1137, and 1115 cm^{-1}, and two components of ν_4, at 667 and 600 cm^{-1}. The water of crystallization absorbs at 3200–3600 cm^{-1} (stretching); 1628 and 1689 cm^{-1} (bending), and 460 cm^{-1} (libration).

Summary and conclusions

As infrared spectrometers were at the time not yet generally available in laboratories concerned with mineralogy work, only nine laboratories submitted spectra, and no more than five spectra were received for any one sample. Nevertheless, the excellent agreement between laboratories

allows the spectra to be accepted with confidence. Such differences as occur
can be ascribed largely to sample preparation technique or selective sampling
(e.g. separation of a fine fraction from the whole sample). However, for one
sample (gypsum) it appeared that the amounts of calcite and silicate contaminant varied among or within the sub-samples supplied to the laboratories.

As expected, the use of the mineral oil mull technique tended to orient
platy particles, whereas in KBr disks these particles were more randomly
oriented. Further grinding was essential to obtain good quality spectra.
With heterogeneous samples, variations in the degree of grinding caused a
variation in the relative intensity of absorption bands due to minor components. The spectra of some samples were strongly affected by their
history. The spectrum of palygorskite is profoundly, although reversibly,
modified by partial dehydration, and gypsum is irreversibly converted to
the hemihydrate above 100 $^\circ$C.

The infrared spectra permit a rapid assessment of the purity and the
degree of crystalline order of the samples. Two samples were obviously
grossly impure: i.e. magnesite, which contains quartz, mica, talc, and
dolomite; and the gypsum, which contains calcite and a silicate contaminant.
Crocidolite and palygorskite contain significant amounts of carbonate
contaminants, and the palygorskite a little quartz in the courser fraction.
Nevertheless, the palygorskite must be considered a valuable member of the
mineral bank, because of its high degree of crystalline order, as indicated
by the sharpness of its absorption bands. The remaining minerals appear
adequately pure, apart from traces of carbonate, and a possibly minor
contaminant in the illite. The illite, montmorillonite, and talc are
shown by their infrared spectra to contain octahedral iron. The illite
is particularly valuable as no mineral of this type has been previously
available in bulk.

References

Layer silicates

Farmer, V.C. (1968), Clay Minerals, $\underline{7}$, 273

Palygorskite

van der Marel, H.W. (1961), Acta Univ. Carol. Geol.Suppl., $\underline{1}$, 23

Ovcharenko, F.D. (1966), Proc. Int. Clay Conf., Jerusalem (L.Heller and A.Weiss, Editors), Vol. $\underline{1}$, p. 299. Israel Program for Scientific Translations, Jerusalem.

Crocidolite

Hodgson, A.A., Freeman, A.G. and Taylor, H.F.W. (1965), Miner. Mag., $\underline{35}$, 5

Gibbsite

Kolesova, V.A. and Ryskin, Ya. I. (1959), Optika Spectrosk., $\underline{7}$, 261-3

Carbonates

Adler, H.H. and Kerr, P.F. (1963), Amer. Miner., $\underline{48}$, 124

Morandat, J., Lorenzelli, V. and Lecomte, J. (1967), J.Phys.(Paris), $\underline{28}$, 152

Gypsum

Ross, S.D. (1962), Spectrochim. Acta, $\underline{18}$, 1575

Wiegel, E. and Kirchner, H.H. (1966), Ber. Deut. Keram. Ges., $\underline{43}$, 718

Seidl , V., Knop, O., and Falk. M. (1967), Canad. J. Chem., $\underline{47}$, 1361

ELECTRON SPIN RESONANCE (ESR)

J. T. Pinnavaia

Electron spin resonance studies direct transitions between electronic
Zeeman levels, i.e. interaction between electronic magnetic moments
and a magnetic field.

CMS RESULTS

ESR spectra were taken for the suite of CMS samples by M.M.Mortland,
Department of Crop and Soil Science, Michigan State University,
East Lansing, Mich. USA.

The samples were converted to the K form and air-dried. The fraction
< 2 ₥m was used. The samples were placed in quartz sample tubes.

Instrumental conditions: Varian E4 spectrometer.

Mode   : derivative

Scan time : 8 minutes

Scan width : 4000 G

Frequency : 9.480 G Hz

Gain   : 3.2×10^2

Modulation : 0.80×10^1

Temperature: ambient

The spectra are collected in the figure:

A – Kaolinite, poorly crystallized (Georgia) KGa-2

B – Kaolinite, well crystallized (Georgia) KGa-1

C – Attapulgite (Florida) PFl-1

D – Synthetic mica-montmorillonite (Baroid) Syn-1

E – Montmorillonite,Wyoming SWy-1

F – Montmorillonite,Texas STx-1

G – Hectorite,California   SHCa-1

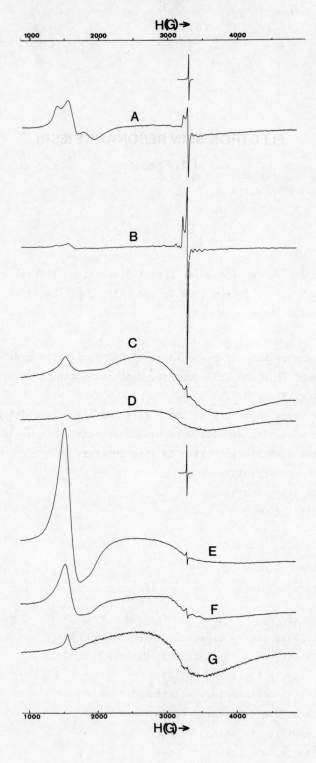

CMS COMMENTS

J.T.Pinnavaia, Department of Chemistry, Michigan State University, East Lansing, Michigan, USA.

The spectra of low and high crystallinity Georgia Kaolinite, spectra A and B respectively, contain two sets of signals centered near $g = 4$ ($H = 1600$ G) and $g = 2$ ($H = 3300$ G). The relative intensity of the low field signal is much larger in the low crystallinity sample than in the sample of high crystallinity. Based on recent observations of Angel et al., 1975, the signals near $g = 4$ almost certainly arise from the isomorphous replacement of paramagnetic Fe^{3+} for Al^{3+}, whereas the higher field signals most likely are due to paramagnetic lattice defects associated with magnesium substitution.

Florida Attapulgite (spectrum C), Baroid Synthetic Mica–Montmorillonite (spectrum D), and the natural smectites (spectra E–F) all show a Fe^{3+} signal near $g = 4$ and a broad line of uncertain origin centered near $g = 2$. The relative intensities of the two lines varies among the different clay samples. The peak–to–peak widths of the $g = 2$ line range from around 500 G in the case of Texas Montmorillonite (spectrum F) to around 1000 G for Florida Attapulgite. In addition to the $g = 4$ resonance and the broad line centered near $g = 2$, Attapulgite and the two Montmorillonite samples show a very sharp resonance near $g = 2$. This sharp line could arise from an organic free radical as an adsorbed impurity.

The utility of Fe^{3+} resonances in monitoring the weathering of clays has recently been reported (Olivier et al., 1975). Also, the shape of the $g = 4$ resonance in montmorillonite has been shown to be useful in deducing the proximity of the interlayer exchange ions to the silicate surface (McBride et al., 1975).

REFERENCES

Angel, B.R., Richards,K.,and Jones,J.P.E. (1975) The synthesis,morphology
  and general properties of kaolinites specifically doped with metallic
  ions and defects generated by irradiation. Proc.International Clay Conf.
  1975, Applied Publishing Ltd.,Wilmette, IL, pp.297-304

McBride,M.C., Mortland,M.M., and Pinnavaia,T.J.(1975) Exchange ion
positions in smectite: Effects on electron spin resonance of structural
iron. Clays Clay Min.,23, 161-166.

Olivier,D., Vedrine,J.C., and Pézerat,H.(1975) Résonance paramagnétique
  électronique du Fe^{3+} dans les argiles altérés artificiellement et
  dans le milieu naturel. Proc. Intern.Clay Conf. 1975, Applied Publishing
  Ltd.,Wilmette,IL, pp.231-238.

AUTHOR INDEX

343